LOST WORLDS ON MARS

Dylan Clearfield

XOX*9099

G. Stempien Publishing Company

Copyright © 2017 Prism Thomas

Editorial offices in New Quay, Cymru

ISBN 978–0930472–21–4

XOXBV657

CONTENTS

Introduction

This book is going to describe the history of civilization on the planet Mars. It is going to do so scientifically, relying primarily upon the discipline of archaeology to present its basic arguments but at the same time with roving imagination assisted by many photographs that provide the evidence. Artifacts and relics from former Martian societies will be on display. The images from the Martian surface will date from the late 19th century to the present. Many of them will be seen for the first time by the public and will not be restricted to the more commonly known pictures of the Martian rover vehicles but will include a wide variety of as yet untapped sources.

IS THIS WHAT PERCIVAL LOWELL SAW BENEATH THE SURFACE?

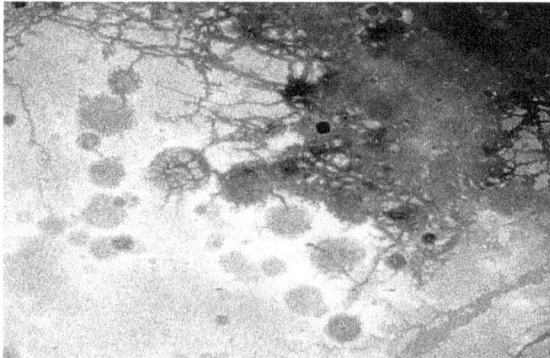

As an experienced archaeologist, I propose studying the planet Mars from the standpoint of an archaeological excavation, using the same techniques, and as such will provide a totally different view from most other investigations done of Mars. By the type of some of the destruction that has been left on the red planet, it appears that at one time in the past this world may have been invaded by a hostile, alien race.

Is this a crashed alien craft on Mars?

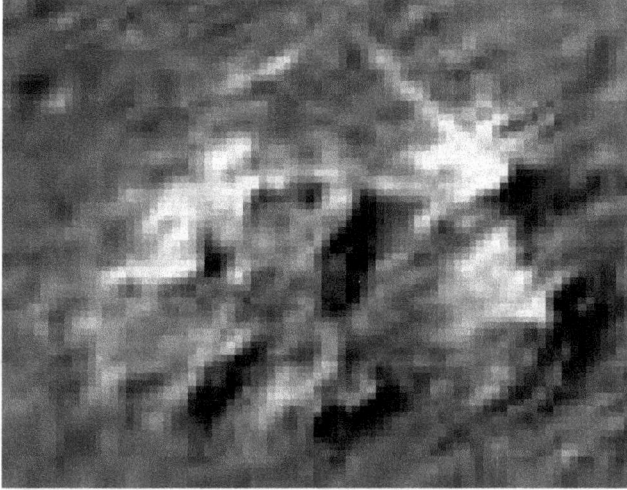

The geological and astronomical history of Mars is very complex. Mars has suffered many cataclysms which have entirely changed the face of the planet and its environment. Its original inhabitants were decimated by a great upheaval which altered the entire ecosystem and which saw them being replaced by a completely different form of intelligent life. The astronomical and geological history of the planet are factual; the social history is speculation based both on the above mentioned facts and the photographic evidence of the remains of former civilizations.

It must be understood that the sites on Mars should be viewed like the ancient archaeological sites on Earth before they were excavated. Most of the sites of ancient Egypt were buried under tons of dust and sand.

They were not simply waiting to be explored in perfect condition. They first had to be discovered. And the way they were discovered was by a trained eye observing a telltale piece of evidence protruding from deep in the sands which alerted him to the possibility of an archaeological site being buried there. That's when excavations began. The same holds true on the surface of Mars today.

Most of the photographs used in this book were taken in the 1970's and 80's by the earliest probes to Mars. As such, they have remained overlooked for decades. Unlike myself, few archaeologists have studied them with the intensity needed to discover those sites which are "hidden" on the red planet under tons of dust. Some of the more recent images will also be used but the older ones provide the key to finding the remains of former civilizations on this planet. Many of these views are dramatic and spectacular and never seen before. This will be a new re-discovery of Mars and its inhabitants.

But is there a hidden danger on Mars? One may be lurking behind the ruins of the seemingly dead planet. The lifeless appearance of the environment is potentially being used by aliens as a type of camouflage. The apparent stillness, an almost suspended state, of this world is a way of luring humankind into a false sense of security. This isn't pure speculation because there have already been hostile acts perpetrated on missions to the red planet. Proof of this will be expanded at the proper time

2. Lost Science

In order to believe that Mars could be inhabited either now or in the past it is first necessary to believe that this planet had the environment necessary to sustain life in some form either in the past or now. That is the purpose of the seemingly technical aspect of this work: to prove that Mars, as Lowell said, could be or have been "The Abode of Life".

It appears that over the years the facts that were known about Mars were deleted or in some way lost to the memory of many of the so-called experts of more current times. What do I mean? Prior to the 1970's many books and research papers were written which described the past geologic and climatic condition of Mars in which the planet was said to have once possessed vast oceans of water. They also reported upon the true conditions of the ice caps and the actual range of temperatures on the planet. These had been correctly estimated by observation even prior to the 1970's. But for some reason, this information seemed to have been lost or was suppressed, making Mars seem to be utterly devoid of water and a barren wasteland from its origin. Even though little of the original data was disproved.

Why would science subvert the facts in such a way? To try to keep the true history of the red planet secret? At whose direction? The answer to this for now is unclear.

The basic facts of the geology and climate history of Mars are known to the scientific community today but it is rare to find anyone who has described them FULLY and accurately to the public. Is it because these facts reveal that life may have long existed on Mars? It's time that this lack of scientific integrity by some experts (NASA) were questioned.

By studying the Martian geology even from afar it has been obvious for decades that water was once – even relatively recently – abundant here. Why did it take NASA several trips to Mars to arrive at this fact and so long to admit it? Or did NASA just not want anyone to know until they couldn't keep the "secret" any longer?

It is a scientifically known fact that Mars has two different types of seasons which have two entirely different types of weather patterns: the yearly cycle and the long range cycle. The long range cycle spans approximately 120,000 years. This difference in seasons makes it likely that at one time Mars was, and will again be, very Earthlike. Why have so few experts taken the extra step to inform the public of this and explain it? There is much discussion about the yearly seasonal change, but none addressing the extreme long range cycle.

Why have we been led to believe that it was a major mystery whether or not this planet ever had large amounts of water on it? It was clear to most planetary geologists and planetary archaeologists like myself that those deep canyons spread across the planet had been made by water. The physical nature of the channels that crisscross the planet reveal that they could not have been made by lava or molten lead or any other type of flowing liquid but water. And even the earliest photographs of Mars revealed what could only be described as islands having once been surrounded by water. Note below from the late 1970's.

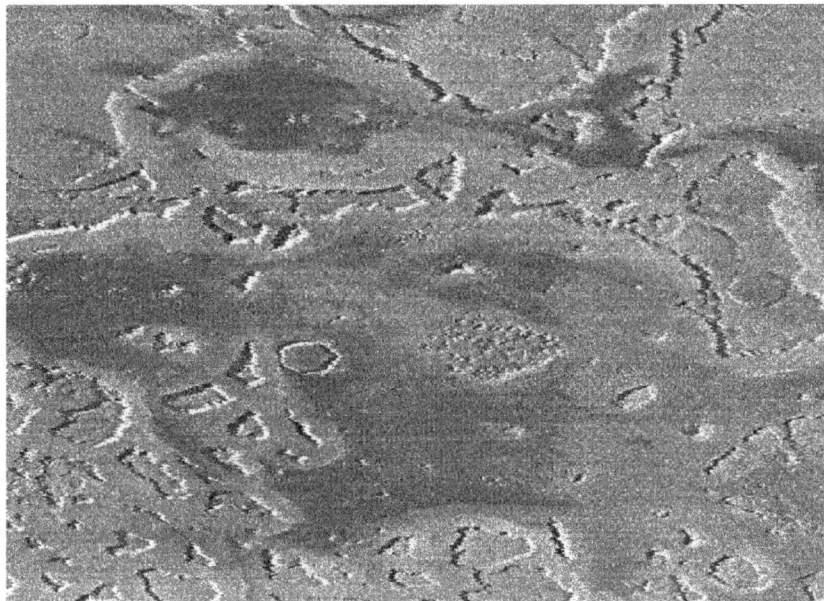

Ismenius Lacus Region on Mars

Why have we been told for so long that this world is barren and uninhabitable when it was known that there is an atmosphere – clearly seen in the variously colored sky – and that weather patterns are active across the planet, implying some form of ecosystem. This is far from a climatically dead world. And this has been known by many free–thinking scientists, including Percival Lowell, for a very long time. Percival Lowell was among the first to proclaim the existence of true H_2O water in the polar caps by noting in the 1890's how the bluish ring that fringed the polar caps after a period of melting had to be the result of trace amounts of H_2O rather than carbolic acid as most had argued.

Many scientists had not even been willing to acknowledge the possible existence of even microbial life on Mars until only recently.

Using common sense, compare the current conditions on the planet Mars with the conditions in the Antarctic on Earth. It has been discovered that microbial life has existed in frozen hibernation for millions of years in the Antarctic and that it had been restored to life after being thawed. Millions of years! Why couldn't the same be true on Mars for the Polar Caps or any other part of the planet? Are some people afraid to acknowledge this for any particular reason? Living beings reviving from hibernation after millions of years! Is this too threatening to some peoples' theology maybe?

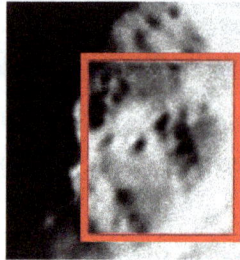

Whose face is this?

But what about human life under conditions of suspension? Could a more advanced form such as this withstand the changes of climate and environment over the extensive 120,000–year cycle, which by the way is a recent climatological occurrence? Compare the long history of evolving hominids on Earth with what might have taken place on Mars. Is the climate during this time span really very different in this regard?

Hominids evolved over millions of years on Earth. Over that time the climate experienced drastic changes, including numerous Ice Ages. Ice Ages when the glaciers spread over a great part of the globe. Did the hominid species die out? No it adapted. In fact, in some cases it grew even tougher and stronger.

Why is this not widely considered as a possible evolutionary history of the Martian race? Why is it easier to say that life could not exist in as harsh an environment as we find on Mars?

3 Crucial Disinformation

Many of the images used in this book came from photographs that are decades old. These pictures were taken by the earlier Viking and later Mariner missions to Mars as well as the earliest orbiters. Not all of the images had been closely examined by the experts but when private study was made of them the pictures were found to contain many anomalies and evidence of past activity on the red planet. Since these images are from a private collection they have escaped the later NASA censorship which was placed upon the official photographs.

While many of the modern photographs of Martian anomalies are trustworthy there are also many images of the Martian surface distributed by NASA that need to be very critically examined before being accepted as genuine. NASA is highly skilled when it comes to dispensing misinformation – particularly about Mars. It will be seen why they do this shortly.

The following photograph of a supposed woman standing on a cliff after strolling from her supposed home on the planet Mars is a prime example. If it looks too good to be true it probably has been faked. And there are certain guidelines to help a person spot a fake. First, distrust any NASA photo that at first appears to be too revealing.

Examine more closely the NASA photograph of the woman who left her house one day for a stroll. The picture of the woman is tantalizingly clear. Conveniently shown in this picture are what appear to be tracks – footprints – of someone leading up to within about ten yards of where the woman is standing. In the background behind her is a house like structure.

NASA has been known to create its own images in order to spread disinformation but whoever does the editing never has been very proficient at it. There seems to be a couple glaring mistakes in the picture of the woman out for a stroll on cliff.

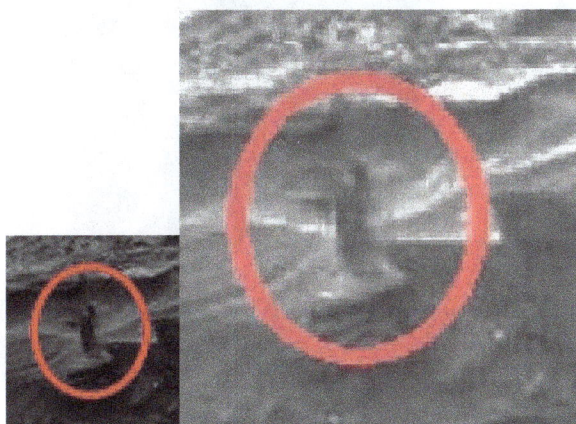

Woman on cliff with a larger view of her on the right

Below is a view of the woman in the red circle, her tracks are in the yellow oval, and the structure from which she apparently came is contained within the blue rectangle.

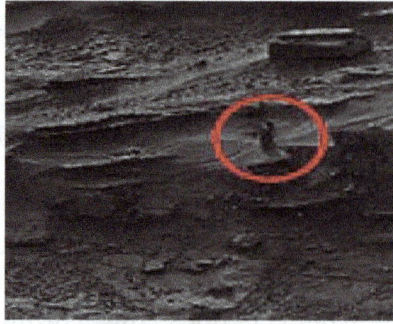

Woman

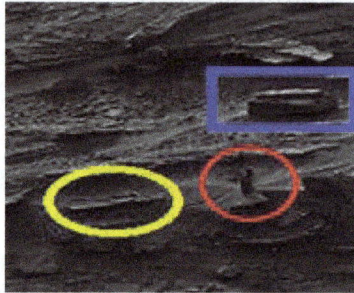

Woman(red), tracks(yellow), and house(blue).

One major problem with this photo is that while it shows the "tracks" made by the woman while on the other side of the ledge it doesn't show the "tracks" she would have left to have made the trek across that adjoining space that leads to the ledge. How did she reach the ledge? Did she float there or did she make one giant leapt for a Martian woman?

Now note the wide angle view of the photograph. Another problem with the photo is the very thin white division line in this picture which is captured within the orange rectangle (also visible in the right side image). It's the place where two images were "spliced" together to form one photograph; and not a precisely matched pair at that.

It looks like a poorly executed piece of editing, not too uncommon for NASA.

There are many new fascinating photographic discoveries being made on Mars as this book is being written. However, other times in the past there were other outpourings of fantastic pictures suddenly being dispensed from Mars by NASA. Some of these may even have been genuine. But NASA frequently downloads breathtaking discoveries from the red planet while simultaneously denying they are of special origin. The reason for this is to spread disinformation.

Here is how the scheme of distributing disinformation operates. First NASA publishes a photograph of tantalizing quality that seems clearly to reveal a startling alien anomaly. Like a Martian woman standing on a cliff. That's what NASA wants people to see and identify: a Martian woman standing on a cliff. But NASA never officially verifies this identification; it simply implies that is what the photograph is revealing.

Then comes the great disappointment. Many months, or more likely a year or longer afterward, NASA decides to perform a detailed photographic high resolution study of the amazing anomaly. The findings reveal that the woman everyone was so excited about was really only a pile of rocks. How many times has NASA perpetrated this trick in the past?

What's the purpose of this? To confuse and condition people to disbelief so that the real truth can be kept hidden. But why hide the real truth? That is something only NASA can answer. What is to be afraid of if there isn't anything there?

However, this tactic of using disinformation failed NASA when it involved examination of the "face." The reason that the "face" became well known is because it was first discovered by independent researchers, namely Richard Hoagland and associates. Then NASA was asked to produce a close up photograph of the location. And NASA did, expecting that the results would deter any more consideration of the "face" being of artificial origin.

Just the opposite happened. It was discovered that the "face," even though currently deformed through wind erosion, was constructed on what was obviously an artificially designed base of immense size. In addition to this, the Cydonia Region became even more prominent. And that brings up the concept of context.

How to determine which NASA photo is the product of disinformation as opposed to one which is genuine? From an archaeological point of view, the test would be context. For example, during studies of the ancient Calico Hills site in California where it was determined that it was settled by a very ancient society about 250,000 years ago one of the primary clues used concerned the context in which the primitive stone tools were found.

They weren't just randomly scattered over the landscape. They were primarily found in specific areas in groups. Not only that, but they were often in various degrees of completion or non–completion if you choose. What this implied is that these were worksites where the tools were being made. This is a common feature in ancient sites, as well as modern sites. More modern examples are kiln sites of later periods where these were located in specific areas because that is how manufacturing is usually done – in specific areas.

This same concept applies to artifacts found on Mars. If something that resembles an iron pipe is found on the surface of Mars in the middle of nowhere without any other similar type of object nearby the best guess would be that it probably is simply something that only looks like a piece of pipe. Why would a piece of pipe randomly be found lying amidst the barren landscape? But if there were a couple of other similar objects nearby then the question becomes: did these pieces of metal belong to part of a building that had been destroyed or were they in some other way related?

The basic method of investigation used in this book follows the same tactic that was used in my earlier work "Alien Threat from the Moon." Images taken of the subject planet are uncovered – in this case Mars – that NASA took decades ago which have revealing objects in them but have either been overlooked, forgot or edited out of existence by the space agency. An archaeological approach is applied to the images. Something that would be of archaeological importance on Earth when viewed from high above would also be of importance on the planet Mars when viewed from high above this planet's atmosphere.

This isn't to imply that the recent discoveries on Mars have been discounted when they are clearly significant, just that they will be viewed much more critically, using the above criteria of context in which they were found. There are some exceptionally intriguing and acceptable newer photographs of objects on Mars as of this writing but do they fit the surroundings or do they seem as if placed there for effect? And remember, someone other than NASA may be placing these objects there.

As a final word: strongly question any explanations that NASA gives to account for any anomaly. Most of their explanations have proved to be quite ludicrous although they claim they are scientific and accurate. As one example is the so–called mystery rock below.

It is circled in red to the right. On the left is the same area just moments before.

This amazing rock suddenly appeared in the path of one of the Rovers after the roaming vehicle had paused. According to NASA, this happened when the rock was run over by the Rover and then flipped over the top of the vehicle and landed directly in front of it. A feat like that would almost be less believable than if an alien or Martian had tossed the rock in front of the Rover.

4 Lowell's Secret

As far in the past as 1898 Percival Lowell saw evidence of intelligent life on Mars. It may not have been Martian life, but it was intelligent. We know this now because it was only recently that his discovery was verified by photographic proof.

Percival Lowell's name is closely linked to the planet Mars. It is also closely linked with the concept of Martian Canals and seasons. His name is also linked with wishful thinking because of this. But his nightly observations of the red planet revealed striking evidence of activity on Mars. While his observations of the so–called canals are well known, no one ever considers his amazing sighting of the heart shaped feature in the Trivium Charontis Region of the planet with only a terrestrial telescope.

Why is this obscure, little known fact of any importance? Because it proves that Percival Lowell was able to accurately see and report upon immensely small and indistinct features on Mars that verify his heightened powers of observation. Lowell's discovery of the heart shaped feature in the Trivium Charontis is such an amazing feat because it should not have been possible to see a feature that remote on a world that far away with any telescope from Earth. It would be as if one could read through binoculars from Earth the serial number written on the side of one of the Apollo 11 lunar landers that had been left behind on the Moon.

How much more plausible does Lowell's observations of canals on Mars then become?

Below is Lowell's own description of the observation of the heart shaped feature in the Trivium Charontis Region taken directly from his personal notes – later published – which were part of his daily viewing notebook:

VIII. ELYSIUM AND ITS SURROUNDINGS.

103. So soon as the daily loss of longitude retrograded its region into view, the Trivium showed. It was at first, June 9, only as a thick canal east and west that I perceived it. The planet was then a long way off and its disk difficultly small; and furthermore its northern position, both by reason of season and of tilt, helped its incognito. In August it could be seen much better. It then appeared as a medium-sized squarish spot, the northern tip of a heart-shaped areola formed of the Cerberus, Tartarus and Mare Cimmerium.

[It then appeared as a medium–sized squarish spot, the northern tip of a heart–shaped areola formed of the Cerberus, Tartarus, and Mare Cimmerium.]

Below is the area on Mars in which Lowell made this sighting. Can you spot the heart?

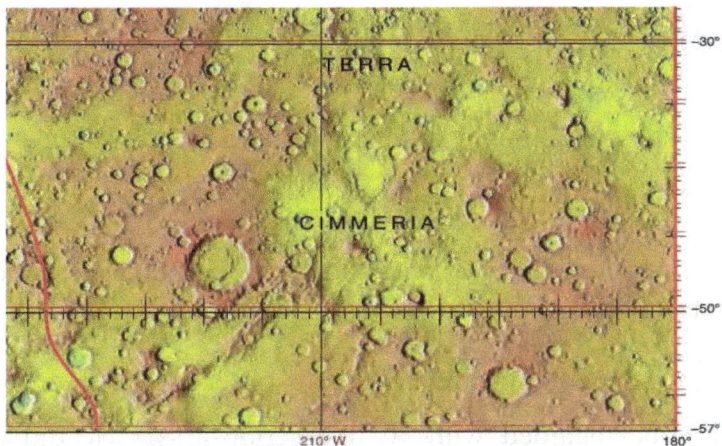

The feature in question is confined within the black square on the right.

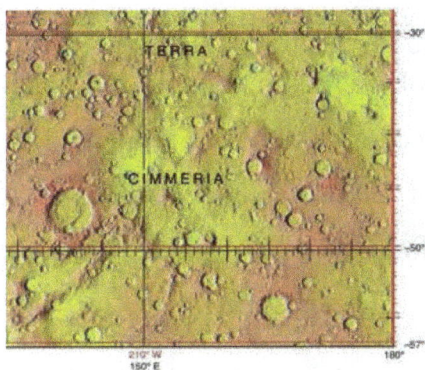 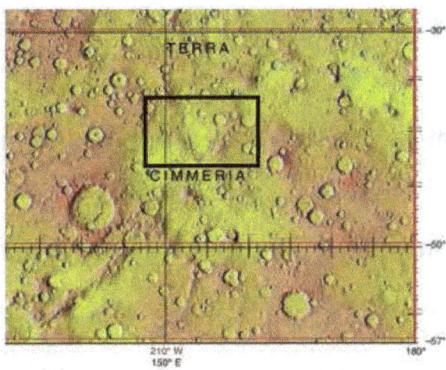

Close up.

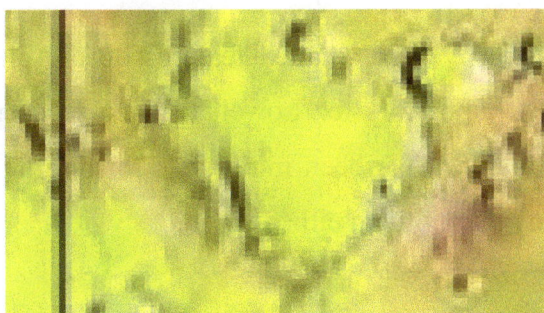

The actual size of the feature is approximately 435 feet across. It is barely visible when photographed from an orbiting craft with the latest technology. How less likely would it have been for Lowell to have seen it from Earth with a terrestrial telescope.

This implies three critical things: 1) Percival Lowell was an observer of the highest quality, 2) the telescope that he used was of extraordinary design and power and, 3) any observations made by Percival Lowell must be regarded with all due seriousness. He did not make outrageous claims based on pure wishful thinking.

In this he might be compared to a very famous archaeologist of the past – Heinrich Schliemann. While everyone who Schliemann spoke to tried to convince him that the city of Troy was a mere myth he refused to believe them and instead, with the book of Iliad by Homer as his guide, went in search of Troy. Using the directions given by the great poet, he discovered the "mythological" city of Troy. Trusting in himself and having belief in a theory, Herr Schliemann made one of the greatest archaeological discoveries in history.

And Lowell persisted in his observations of Mars and believed in his vision.

There are many heart shaped features on the planet Mars but it should have been impossible for Lowell to have seen any one of them. And, as such, it should have been impossible for him to exactly pinpoint the location of any one of them either.

There is an interesting irony connected with the discovery of the heart shaped formation on Mars. This concerns the major planet Pluto. Percival Lowell predicted the existence of Pluto even to the extent of calculating its position. His student Clyde Tombaugh discovered this major world of our solar system in 1930 based on Lowell's formula. How curious is it then that one of the most prominent features on Pluto is a heart shaped area in its southern regions?

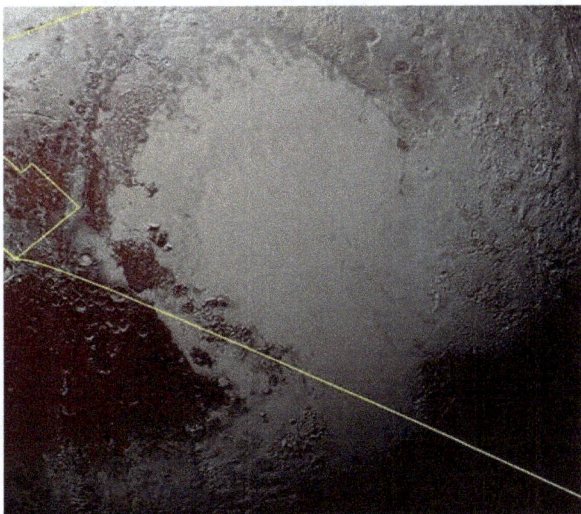

Coincidence?

Did Percival Lowell really see actual canals on the surface of Mars? Probably not. But the critical word here is SURFACE of Mars. The question should be rephrased as: Did Percival Lowell really see the covered, below ground remains of what may have been canals or something similar?

Lowell at his telescope

Lowell used an *Alvan Clark & Sons* 61 centimeter (24 inch) refracting telescope to make his observations of Mars from Mars Hill, Arizona. This specific telescope was renowned for its superlative quality and uniqueness in regard to ability to magnify objects in the distance. It should also be noted that being a refracting telescope rather than a reflecting telescope it was naturally much better suited for planetary observations.

Were those really canal's that Lowell saw on Mars? Many people probably consider this a ludicrous question to ask after all the images sent back from Mars by orbiting craft. But what inspired Lowell to draw the maps seen below? Wishful thinking or exceptional observational ability?

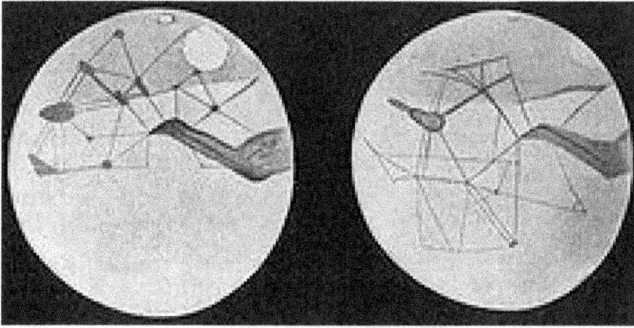

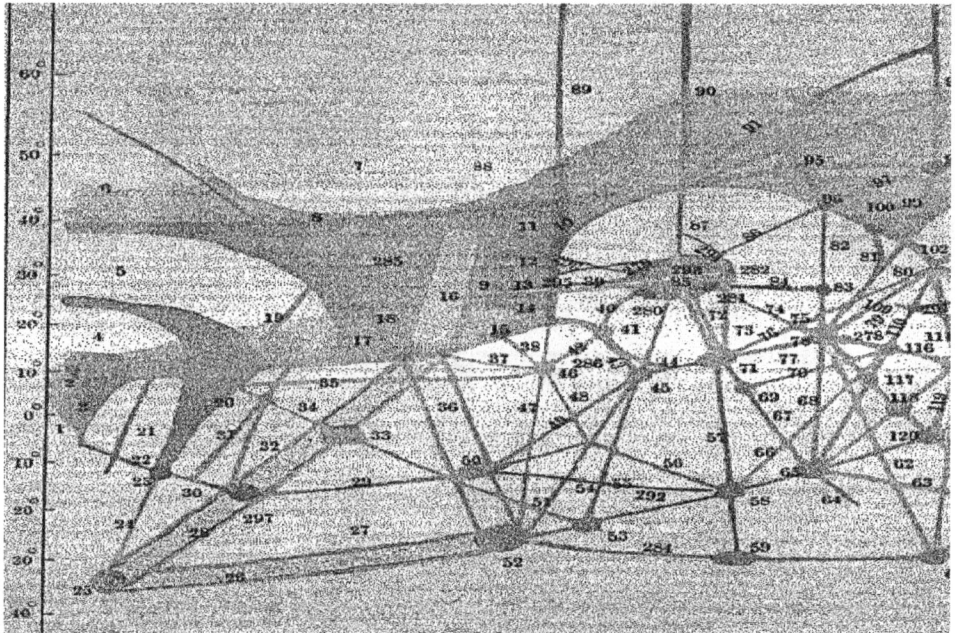

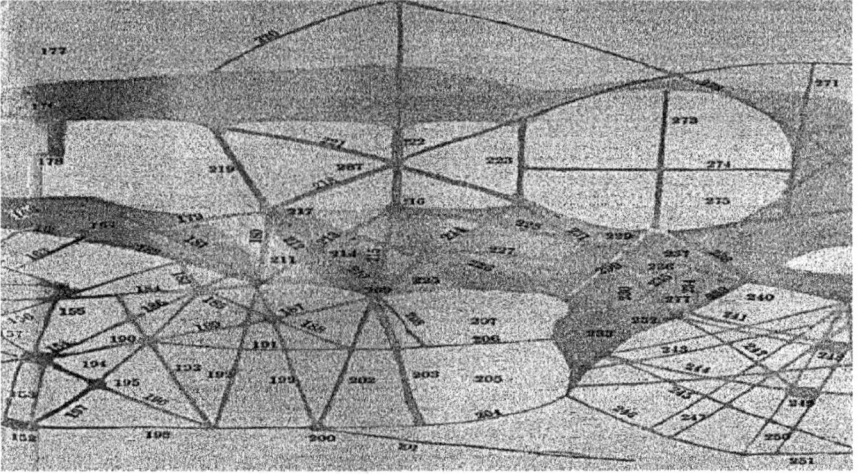

Note the images that follow. These are what it looks like beneath the upper surface of Mars in photographs taken by the MOLA orbiter. Could something like this be what Lowell saw buried under the layers of thick soil and dust?

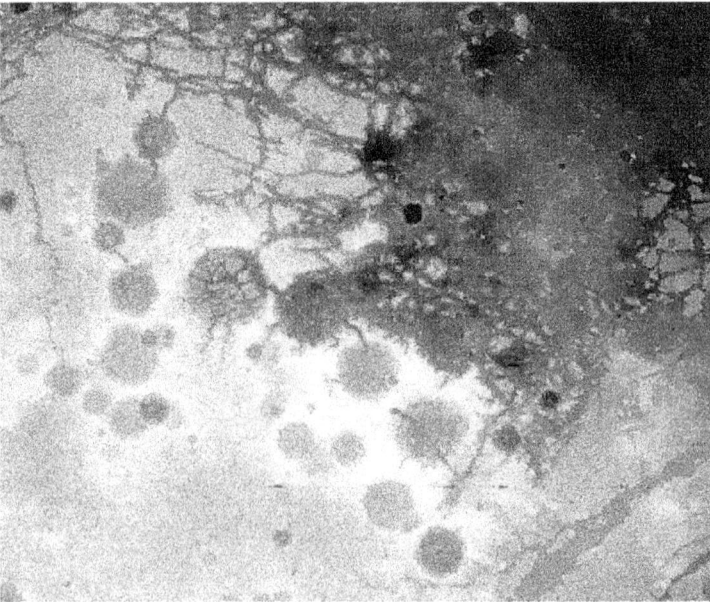

30

Plate 1

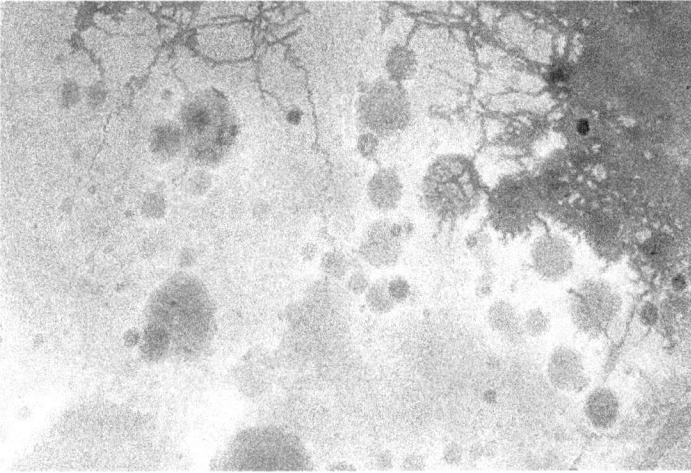

Plate 2

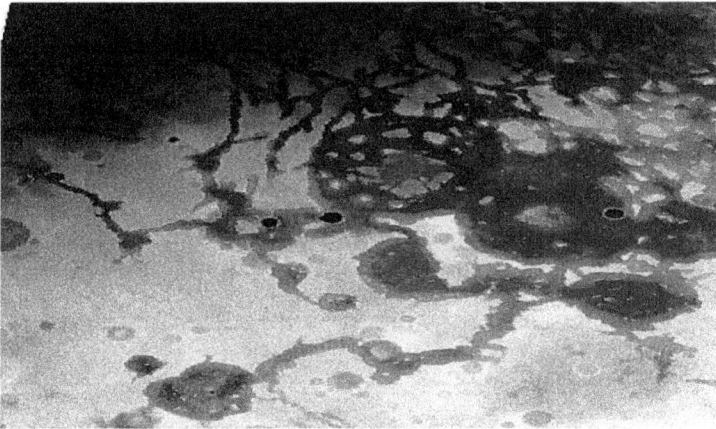

Plate 3

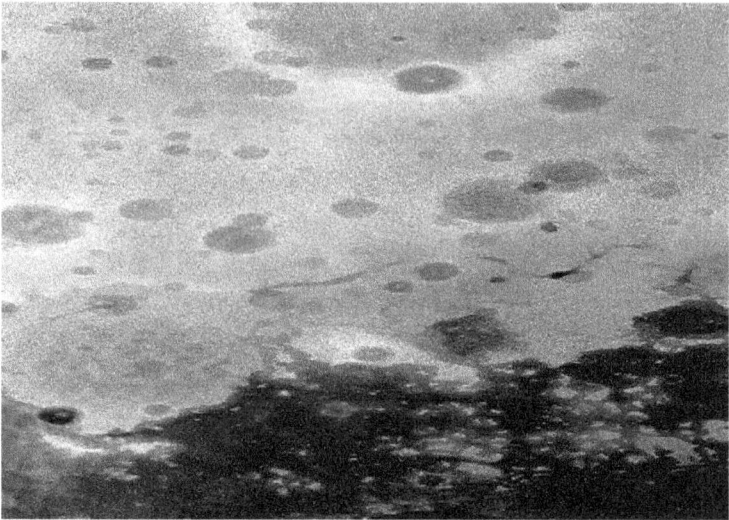

Plate 4

These do look like canals and accompanying oases, do they not? Is it just possible that this is what Percival Lowell was able to see? These are images publicly available from: USGS Public Explorer maps. As already noted, they are from the MOLA project (Mars Orbital Laser Altimeter). The light areas show higher elevation; the dark areas are primarily subsurface.

While these may or may not be canals, what other objects might they be? Two suggestions immediately come to mind. Could this network of interconnected lanes and "oases" be transportation tubes like subway systems belonging to underground urban centers which are thus connected together? Or could this crisscrossing matrix be some form of electrical power grid or a transporter of some other form of energy which was, or is still, spread across the planet below the surface level by extra–terrestrials who have colonized this world? Why isn't this possible?

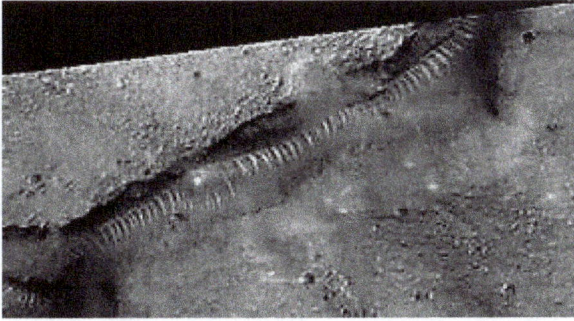

Underground energy line or natural feature?

Another suggestion concerns the possibility of these forms being organic and as such alive. Could these in fact be vast areas of living lichen spread across the red planet adding to its crimson color?

Something to consider. Especially when looking at this next photo of what most people believe is simply an indentation in the middle of a crater on Mars. But take a closer look at this so–called indentation.

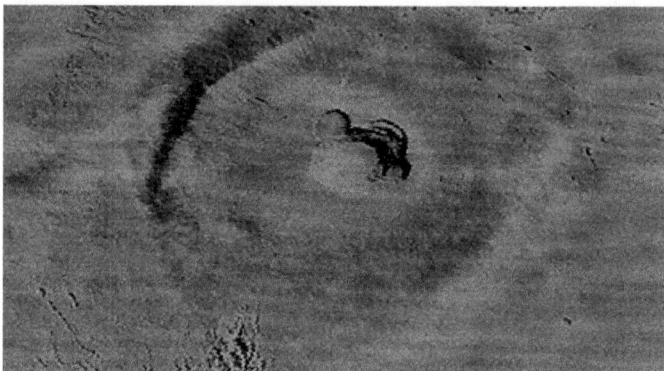

Crater?

It looks a lot like this doesn't it:

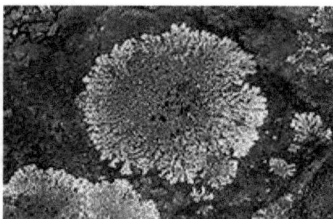

Could it also be an intelligent form of lichen? And does it change color from one yearly Martian season to another to account for the vegetation pattern on the red planet?

It would be ironic if the life on Mars should turn out to be vegetable in nature. In the 1951 original film "The Thing from Another World" (shortened to "The Thing") the monster in the movie was actually a gigantic vegetable. And as such it was in the Plant family and as such it required carbon dioxide to breathe for photosynthesis to take place and the atmosphere of Mars is mostly carbon dioxide. Coincidence?

No matter what this formation just below the surface may be one thing is certain: it is there and it has been photographed.

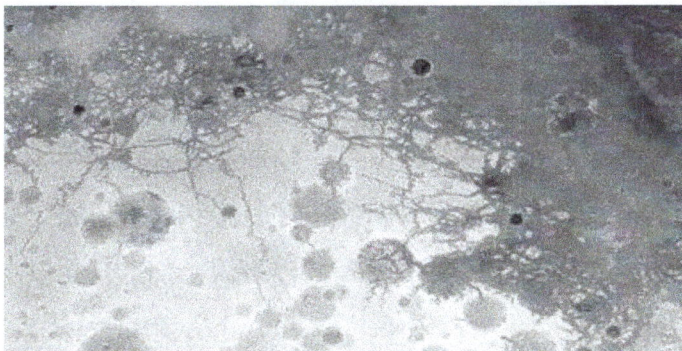

And what Lowell saw.

Lowell made another astonishing discovery in the early 1900's that verifies modern day evidence that aliens are probably on Mars right now. This has already been alluded to and will be fully examined later.

5 Mars of Old

As noted earlier in this book in order to believe that Mars could be inhabited either now or in the past it is first necessary to believe this planet could in some form possess or have possessed the environment that could have or could now sustain life in any form. That is the reason this brief chapter is so important. It will demonstrate how Mars in the past could have been the home of many types of living beings and even today could harbor life.

Many millions of years ago, Mars was quite a different planet. The lake basins would have been filled with water. Vegetation would have been in bloom. Great oceans covered great areas. And there were great oceans.

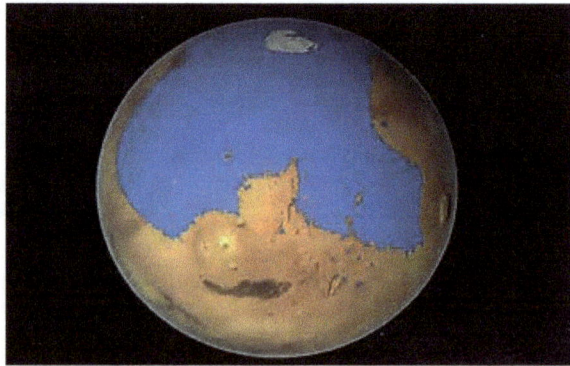

The atmosphere would've been thicker, having been released from suppression within the soil and rocks. The air pressure would likewise be more sustainable for life. And there may have been Martian inhabitants of a high order engaged in activity while on the Earth at this time Homo Erectus was beginning to rise on his two feet.

It is speculated that about 65 million years ago Mars was enjoying a favorable obliquity to allow for a rather pleasant climate, very similar to what we currently have on Earth today. Obliquity is the key. It refers to the degree of a planet's tilt on its axis. Mars has a very eccentric obliquity – range of tilting – now which accounts for its epoch length seasons. But in times long past the tilt of Mars had been much like Earth's.

The tilt of the Earth on its axis – obliquity – plus other factors is a major reason that accounts for our yearly seasons. Because Earth's obliquity is relatively stable, ranging between 21 and 24.5 degrees, the planet enjoys a relatively stable yearly climate. Even the slightest change would make a great difference. Some scientists believe that a simple change in tilt of one degree was the cause of the last Ice Age. That's how important obliquity is.

Consider that Mars now has a range of obliquity of from 0 degrees to 60 degrees and it can be imagined the effects this would have on a planet over the passage of an epoch.

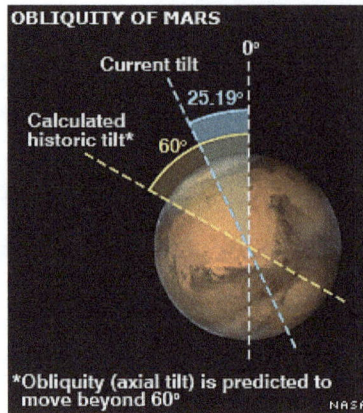

OBLIQUITY OF MARS

Current tilt — 0°
25.19°

Calculated historic tilt*
60°

*Obliquity (axial tilt) is predicted to move beyond 60°

NASA

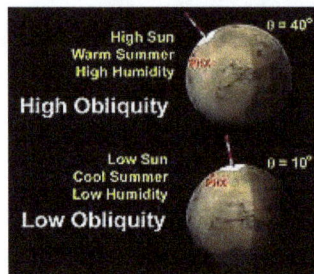

High Sun
Warm Summer
High Humidity
High Obliquity
θ = 40°

Low Sun
Cool Summer
Low Humidity
Low Obliquity
θ = 10°

Mars now has a cyclical season that spans about 120,000 years. It actually takes a great deal of time for the wide degree of change in the planet's axis from 0 to 60 degrees. This means that for about half of this period – 60,000 years – the Martian environment would probably be sustainable for life as we know it. But what about the other 60,000 years? Underground cities? Hibernation? Flight to another world? Leaving the planet doesn't seem likely for two reasons. First, if the Martians were that advanced they could at least partially terraform their world. Secondly, the Martians probably never advanced far beyond a mid– 20th century type of society. Both of these reasons would also eliminate the idea of highly developed underground cities.

This leaves some form of hibernation as being the most likely alternative if Martian civilization had survived the great catastrophe to be considered later in this work.

Some find it difficult to understand how Mars at this period in time can have two different types of seasons. How can there be a yearly season – lasting about 685 days – and yet also a cyclical one lasting about 30,000 years each season? This is seldom explained and it is critically important to know.

Here is the explanation. During each revolution of the sun the tilt of Mars changes by a minute amount – a single percent. Thus, the seasonal changes that occur during a yearly span are not that great. But with the continued tilt in the axis changing little by little over a thousand years in time a difference in climate becomes noticeable. It is an extremely gradual action. The real – epoch length– seasons of Mars change at a very slow rate, but they do change. Winter becomes spring and spring becomes summer. And yes, there is even an autumn.

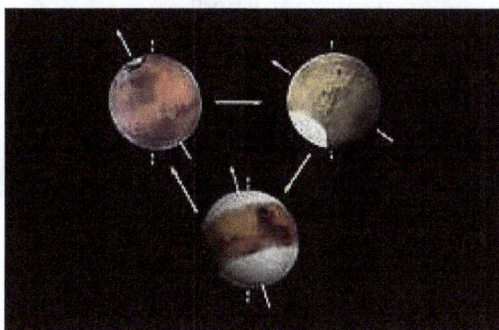

Consider the type of environment with which a native Martian would be presented. Perhaps a ten degree change in temperature

Fahrenheit each century due to the intensely slow change of obliquity. That could mean that for each passage of a century a slightly newly evolved species of Martian would develop in order to acclimate to the changing environment. After 1000 years a different type of Martian would have evolved, but not so different as to be unrecognizable to his forbearers. In other words, the degree of change in the climate would provide a gradual enough transition to make a gradual evolution possible and sustainable for eons.

Martians might have even dated their calendars according to the stage of obliquity that exited during their time on the planet. Entire historical events might have been recorded in this way. For example: during the age of 46% obliquity the southern polar regions were first settled.

Using Earth climate as an example, the great Ice Ages could be viewed as like the winters on Mars. Humans – primarily Neandertals – existed and survived through Ice Ages. Why couldn't Martians do the same on their world when winter returned?

What about the water? Would that come and go, disappearing into the subsurface and becoming trapped in the rocks or at the poles? Probably. It has been proved to have done so during the current age thus it probably would have in the past.

Maybe, as will be shown, the original Martians were able to artificially store water below the surface in huge reservoirs accessible through an intricate system of pipes. But all of this only matters if the original race survived the great catastrophe, or if it was afterward that Martian civilization first developed.

That isn't the premise of this book. It is speculated that the original Martians were annihilated by a horrific natural disaster and it was after this event that the climate became so hostile. That is when the aliens arrived.

6   The Great Eye

Despite all of the technical hardware that has been sent to Mars, the red planet is still a land of mystery and enchantment. This chapter will examine some of the more curious, and well–known, features on the planet. But it will begin with one of the least considered areas on Mars even though it has a very pronounced presence: the great eye.

Few people have even considered this feature or know what it is. Even fewer still are aware of its bizarre relationship with the pyramids of Mars.

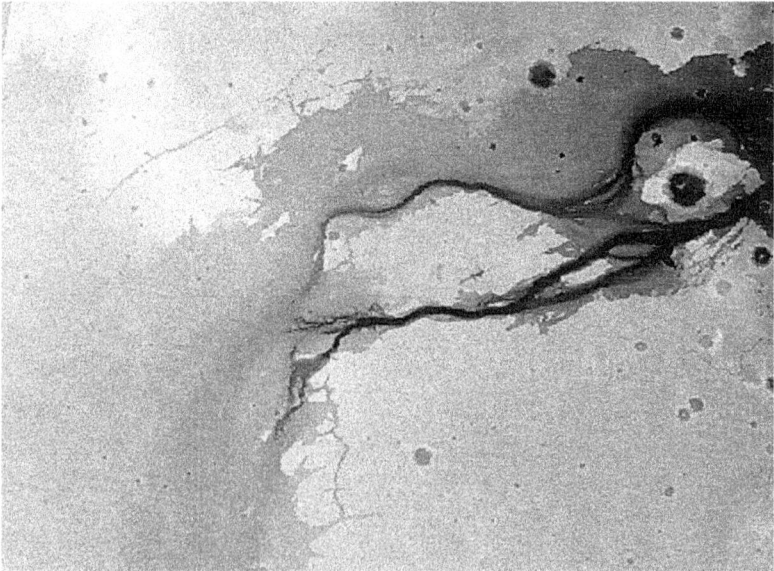

The great eye

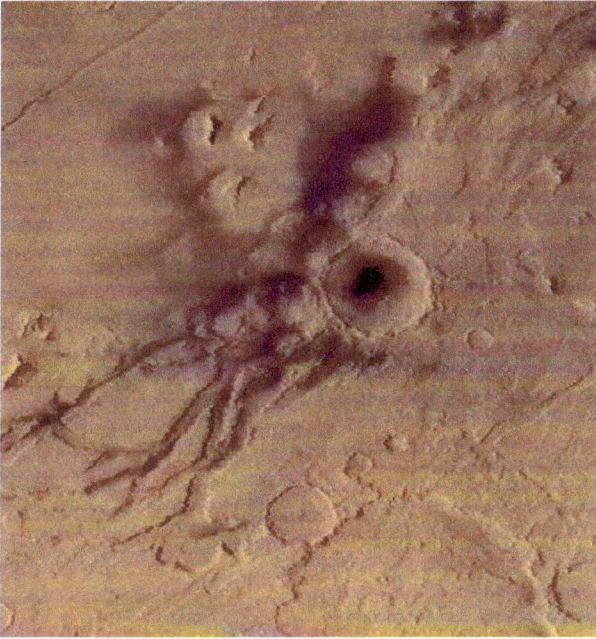

Great Eye different view.

The great eye is known as Lucus Lunae (Lowell) and is a natural feature on the planet. As an anomaly its interest lay in an implied relationship between it and the pyramids, a relationship which is only a curiosity. The relationship mimics the well–known image of the All Seeing Eye that is often found in symbolic fashion at the apex of a pyramid. It is also found on the back of a U.S. dollar bill.

ALL SEEING EYE

The All Seeing Eye was used in ancient Egypt and was known as the Eye of Horus. It was a symbol of protection and good fortune. In other uses, the All Seeing Eye is meant to represent the power of God which is always looking over his creation. And in the United States the eye is at the apex of a pyramid whose symbolism is meant to imply seeking the aid of the architect of the universe over the nation's endeavors. In some cultures, however, this eye is meant to represent an evil presence keeping eternal watch over you. Is it therefore somewhat strange that the great eye on Mars should be found in near association with the pyramids of the red planet?

Many types of pyramids have been found on Mars, the most famous being in the Cydonia region. Objects of pyramidal shape are more likely to be the natural product of human – or Martian – engineering than that of nature. Like the pyramids of Egypt, Peru, Mexico and other places, the structures on Mars would have been one of the most natural designs to be created by a society of intelligent beings. Parallel development.

But could pyramidal shapes be artificially constructed on another planet? Yes. It's called independent invention. But, like on Earth, might the construction of pyramids on the red planet imply more than an artistic statement? Might not the pyramids on Mars serve a scientific as well as cultural purpose, perhaps by being part of a communication system – maybe even one that has contact with the pyramids on Earth? Or on our Moon where pyramids have also been discovered?

Also consider that on Mars, just as on Earth, the pyramids might also be of use to a higher civilization. On Earth, communication towers have been placed atop ancient pyramids. It may be possible that on Mars aliens who have colonized the planet are using the pyramids for their own purposes. Maybe even as energy generators.

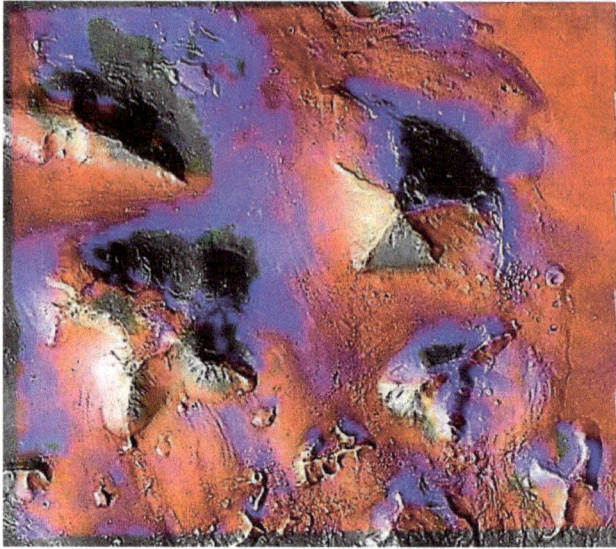

But are these edifices truly made by Martians or simply the product of wind and erosion? There is yet to be discovered a pyramidal form anywhere – Mars or on Earth – with a totally smooth surface on an angular plane having been the result of wind or erosion. It isn't something that naturally occurs. It just is not!

46

Two or more smooth angular sides. How?

And if this were the product of erosion, why would the pyramid still keep its shape? Would it not be more likely to have been worn down into an indefinite form and be much more lopsided? After 65 million years? That is the time frame this report is assigning to the "face," the nearby pyramids in Cydonia, as well as to many other locations on the planet. If dinosaur remains still exist so too can the remains of huge rocky formations in a cold, dry climate.

While the sides may not be perfectly smooth on the pyramid above, the ridges at its apex are not the usual result of mere erosion. How would they have formed in such a sharp edge with a point on the top? Maybe that point on the top is something more than stone.

Most people compare the Martian pyramids to those in Egypt and sometimes to those in Mexico or on the Mayan Peninsula. However, they seem to more closely resemble those which are found in the area of Nazca Peru, more specifically those in Cahuahchi. Note the photos that follow:

This was a pyramid complex constructed about 1,500 years ago by the ancient Nazca. It is mostly buried in soil and sand. The covering of sand was both intentional – for reasons as yet unknown – and caused naturally by the wind (similar to Martian wind power?). At the base of the pyramid is the ritual city which supplied homes for the priesthood. A stylized view follows.

This dusty, wind driven barren landscape in Peru seems to be a much similar match to that which exists on Mars now which preserves the relics.

The following series of images is taken from the Elysium Region of mars. Not only does it reveal a group of pyramids but other features that are startling in consequence. The photo is from an early orbiter craft.

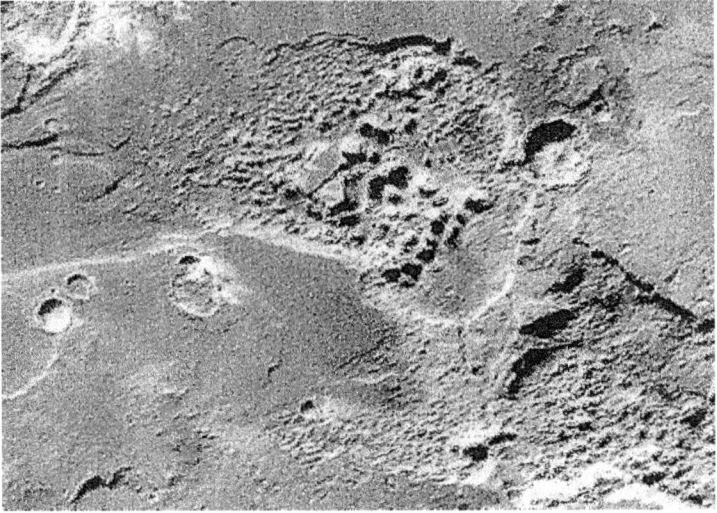

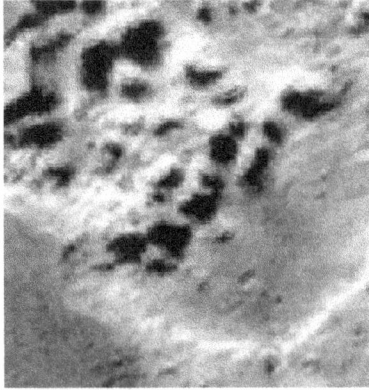

Pyramids at mid bottom

There is much more than a group of pyramids in this picture. Note below where the top arrow points toward what looks like an obelisk or monolith. The sideways facing arrow is directed at a very angular S–shaped wall. The pyramidal structures are within the circle. These features are too regular of appearance to be naturally formed. If these were on Earth and spotted from overhead, they would warrant archaeological investigation. And I would know because I have done so.

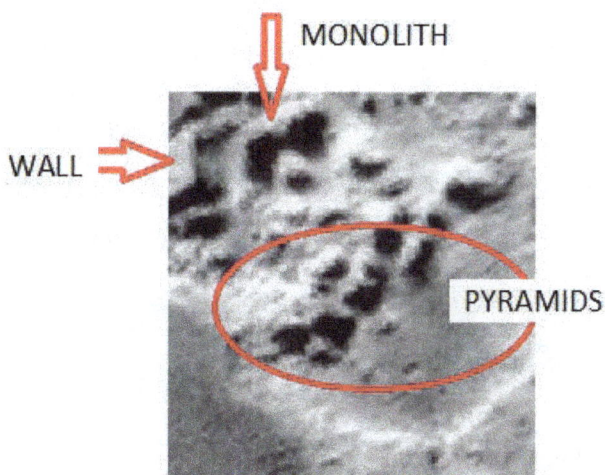

There is one exceptionally important feature to notice on one of the pyramids. At first it was suspected to be an artifact of the photographic process. But on closer examination it seemed to be an artificial attribute on the apex, or top peak, of one of the pyramids. Any further enlargement would wash it out of existence so the reader is asked to focus on the area below within the yellow square and note the Y–shaped feature atop the pyramid.

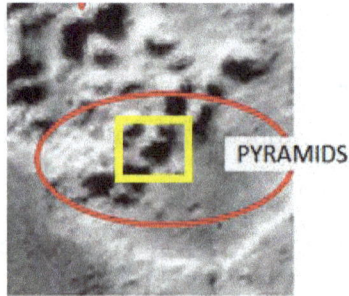

PYRAMIDS

A star shaped figure atop the pyramid in yellow square?

Notice this same area from the original normal size photograph which has not been tainted by any photographic artifact. The star shaped feature can still be seen. Could it be a form of antenna?

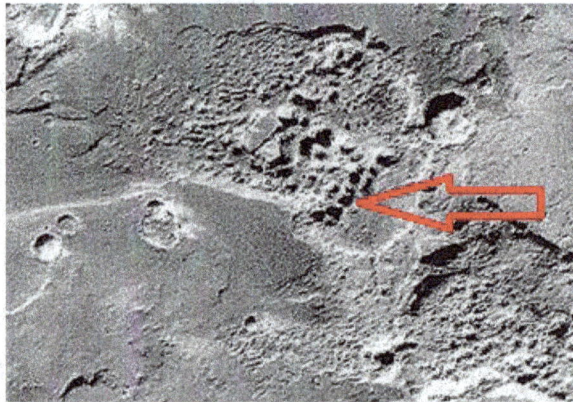

What follows may be one of the sharpest images of a building on the entire planet. But you do have to use a little imagination. Consider this as a cluster of buildings rather than a grouping of stones of almost impossible size. Solitary stone blocks in this shape and the size of huge buildings do not normally occur in nature.

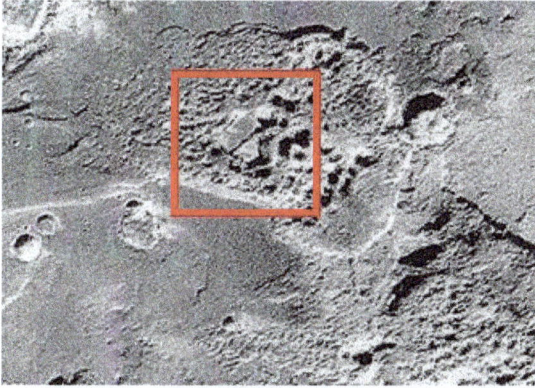

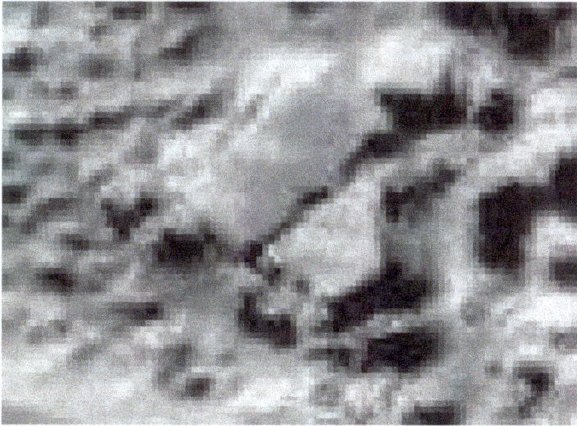

Enlarged

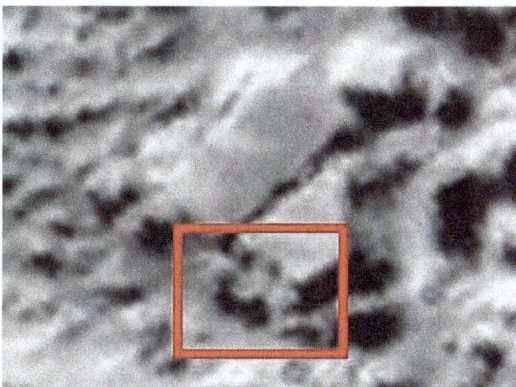

Entrance to the building with overhang!

There is still another feature of this complex to be examined. And it is just as fascinating and intriguing. It is directly across and to the right of the building that was just investigated and it appears to be part of a type of military compound.

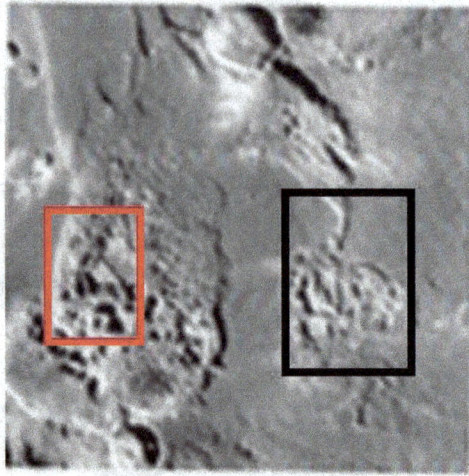

Building with overhang in red. Military compound in black.

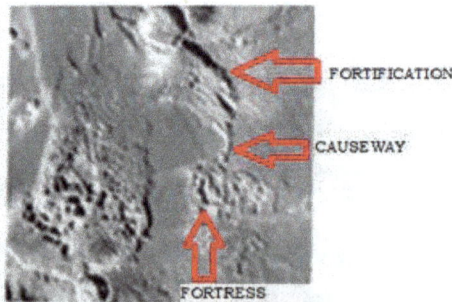

Military compound specifics.

A person could ask: why are there pyramids associated with a modern type of building and a fortress of 18th century type architecture? That is a reasonable question. A structure that has a pyramidal shape is not necessarily a pyramid in the ancient Egyptian sense. There are many modern buildings with that shape or something similar to it. Thus, the pyramids being examined in this context may be of the same vintage of the other buildings nearby or they are of older construction but being refashioned for a modern use.

The proximity of the fortress to the more modern, mid 1950's type building can be explained as being a still useful piece of architecture existing alongside a then state of the art structure. This isn't archaeologically unusual. No matter how this complex is viewed, it provides a rare insight into Martian history.

Archaeologists usually perform their initial observations of a potential site from above because that is where the best view is had. On Mars the Face was a creation that was meant to be seen from above. On the Nazca Plain the images were also meant to be seen from above. Were they both signaling the same observers? Below is the Cydonia Region.

The Egyptians, the Mayans and the people of Nazca were all proficient at mummification. It can at this time only be wondered if mummies will also be found on the planet Mars.

Some familiar images are about to be re–examined in a very new way. Remember, the main focal point of the arguments in this report is common sense supported by basic scientific reasoning and archaeological comparison.

The first item has been investigated from almost every possible angle imaginable. So let's look at it from a different angle, one which a planetary archaeologist would use.

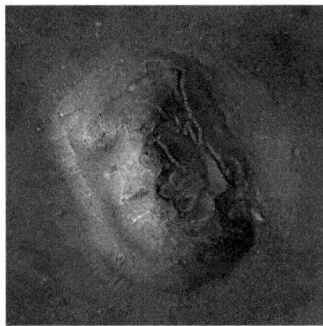

It doesn't look much like a face from this lower vantage point in orbit, does it? Yet, it does if you look at it from the next vantage point which is much higher.

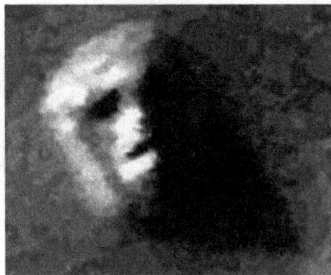

And isn't this the vantage point from which it was meant to be seen? Like the images on Nazca Plain in Peru the face was a symbol that was directed skyward at someone who would be passing by overhead. Or maybe it was meant to be seen by someone on one of the two Martian moons. Phobos and Deimos may have truly seemed like gods if, as suspected, they mysteriously appeared from nowhere in the Martian sky by a method as yet unknown. However, it can only be guessed at who the creators of the Face had created it for.

But would the Martians purposefully create a distorted figure and hope that it would look like a full featured face from above?

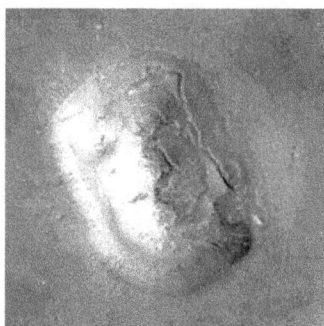

Probably not. It has almost surely been worn down by wind and other erosional factors. But whoever constructed it may have had the foreknowledge to realize that the face that they created would over time be gradually worn down by erosion. They would expect this. And might they not have been smart enough to be able to design this face so that even if it was worn down by erosion its primary features would still remain relatively intact if seen from high above or even in orbit?

It's a concept similar to the one used by the ancient Greeks when they purposefully built the columns of the Parthenon slightly at an angle, realizing that if seen from the distance the effect would make the columns appear to be perfectly straight.

Even if the idea did not occur to the Martians that a partially eroded face would still be recognizable as a face from above it is still easy to visualize that the form of a face once existed. In other words: there may not be a fully formed face there now – **BUT AT ONE TIME THERE WAS!!** And it didn't form by itself.

That is just part of the argument. Here is another angle from which to observe this face. Consider the base on which the face stands as mentioned earlier. The base is a roughly rectangular structure. Even as it exists today the rectangular base still survives. As an archaeologist – and it doesn't matter which world is being examined – a base of material that is this large and this uniform does not form by nature. It is formed by intelligent design. How would the howling winds of Mars create such a base on which the face is supported?

From an archaeological point–of–view the placement of the face on the outskirts of this cluster of pyramids is precisely what would be expected. In a way it is reminiscent of the Egyptian Sphinx and may have served the same ritualistic and magical purpose.

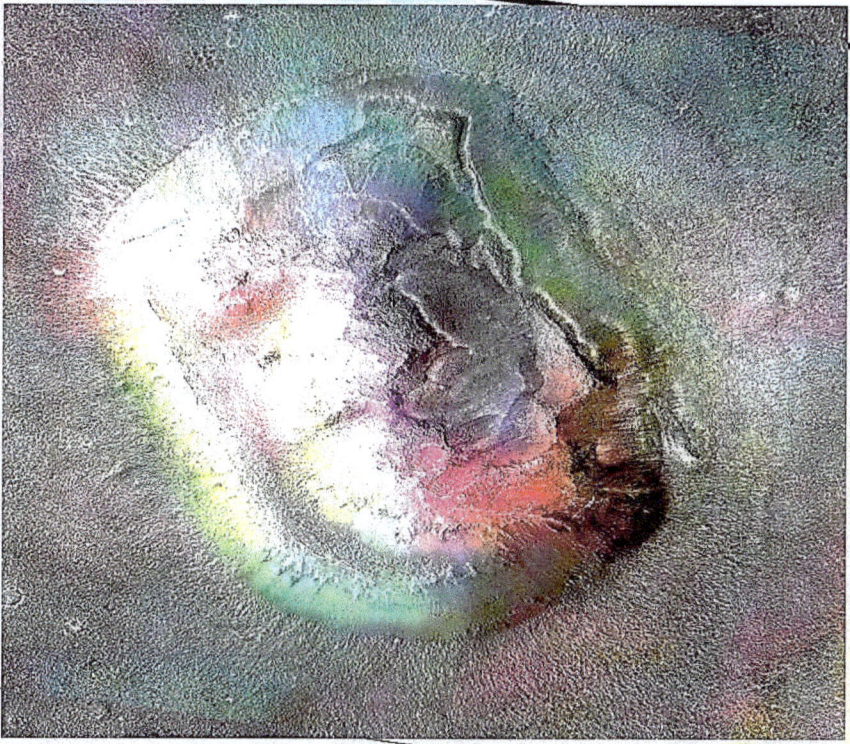

Even in extreme close up it is obvious that the base upon which the Face is placed was artificially made.

Like in ancient Egypt, there seems to be the threat of a curse attached to the monuments of Mars. It may not be scientific or provable, but perhaps the features of the great eye and the face and the artifacts at other sites on the red planet are not signs of welcome but of warning.

7 Bottled Water on Mars?

When considering life on Mars during any epoch the question that always arises concerns water. Was there water? How much water? And where did that water go?

As noted earlier, it's been known for a long time that water once existed on the red planet. Where is the water now and how long ago has it been since the last major supply of water disappeared from the surface? And before the water disappeared how did the Martians conserve and distribute it?

The answer to one of the above questions appears to be that it may have been as recently as several million years ago that there was a substantial supply of water on the planet's surface. This is implied by the photograph taken by one of the Rovers of an object that is clearly related to water storage facilities.

This is the so–called Y–valve (even though it looks more like an X).

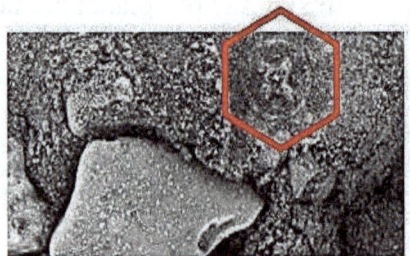

Not only is it an impressive feature by its design but maybe even more so by its appropriateness. These types of valves control the suppression or the release of water or other liquids. This is the type of device that would be needed to keep water supplies under control and for preservation. Below is what an operational one with fittings looks like.

The image of the Y valve by itself on the Martian surface is impressive but it is made even more so by the circular object that appears directly to its left.

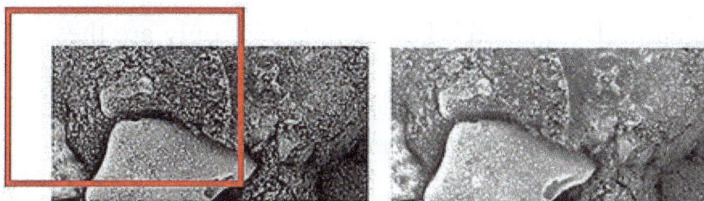

There appears to be a circular, possibly metallic, plate lying next to it and partially covered by a very large rock. And if you look even more closely at the top of the plate a type of flange or latch can be seen (below in red).

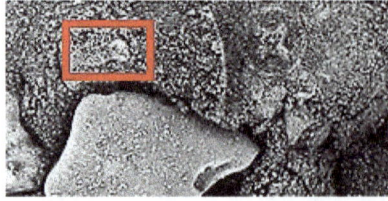

And a larger version of the key–shaped flange:

Can all of these features taken in context be coincidence? Can any of them be naturally created? I think not. Not only are they not of natural origin, the detail involved makes this even less likely. Every specific feature that should normally appear on a Y valve of this type is there where it should be in the form that it should take. Below – an earth–based top bolt.

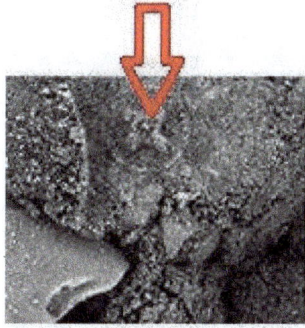

A Martian top bolt?

Far beyond coincidence! And maybe what Lowell saw and thought were canals was really a vast system of underground pipes that delivered water across the planet. How did he see them if they were underground? Maybe it could be explained simply by the type of aperture he used on his telescope which could genuinely account for his unusual viewing abilities. Or maybe his unusual viewing was allowed by a specific filter he used.

Compare the following two images:

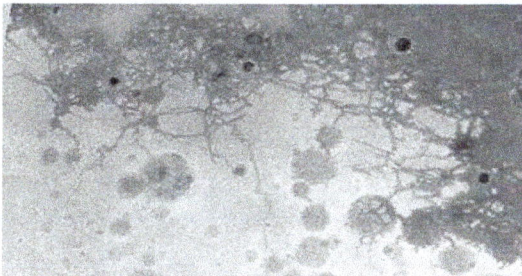

What Lowell saw on Mars

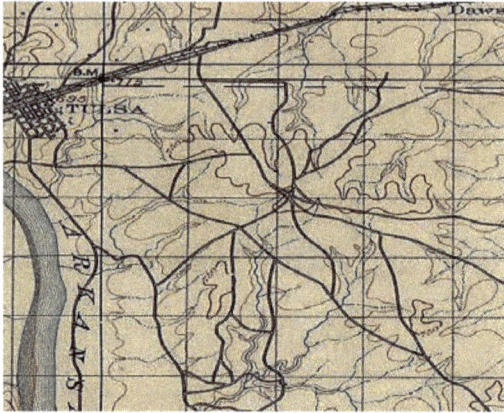

Underground pipe system in Oklahoma

The forms are similar, not exact. Perhaps the Martians or whoever had colonized Mars had a different system of pipe laying from those used on Earth. But the similarity is close enough to make a comparison of usages – transporting water over long distances beneath the ground.

Also compare this diagram of a natural underground water supply on Earth.

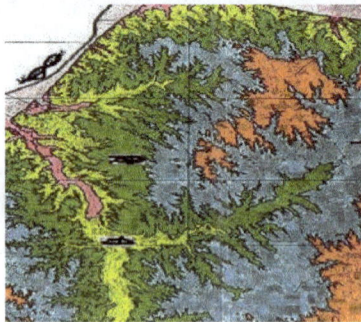

Underground water sources

Maybe something like this exists under the surface on Mars, too? As of this writing, NASA is preparing to send a craft to the surface of Mars specifically to dig below the ground in search of subterranean water sources. Hopefully, they will dig near the location of the Y, or X valve pictured above.

Most people aren't aware that rivers on Earth that have become totally dry on the surface often maintain a supply of water under the surface.

Yes, water might exist beneath this dry river shown above. One needs to dig deep enough but more importantly one needs to know where to dig.

On the Earth there are above ground storage facilities for the water. Below is an example of one of them in the process of being constructed. Only the roof remains to be finished.

Note the next photograph below that was taken of the surface of Mars from a 1980's orbiter. Notice any similarity between it and the water storage system on Earth?

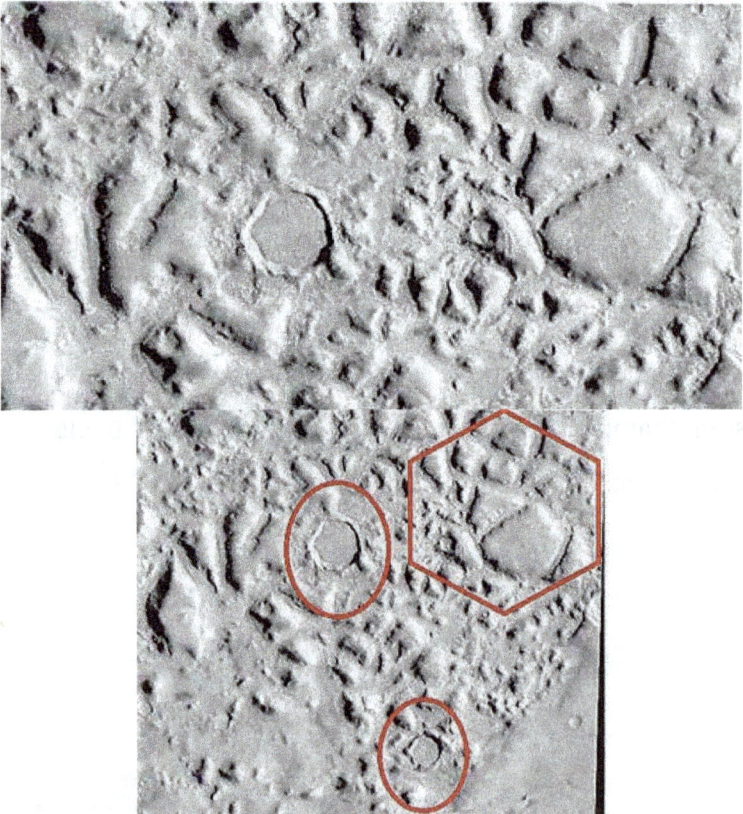

Are these really only natural features?

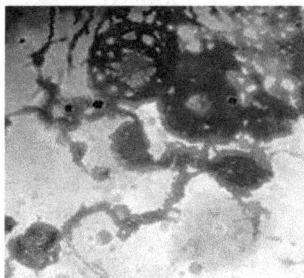

Are all of these just coincidences of design? Are those formations in the Acidalium Region pictured above only natural features, too? How many of these similarities can NASA continue to argue are simply coincidences?

It is suggested here based on archaeological comparisons to earthly designs that the water storage facilities are what Lowell described as the Oases on Mars to which the canals were connected.

Also of note are photographs on Mars of what can be nothing other than dried out reservoirs:

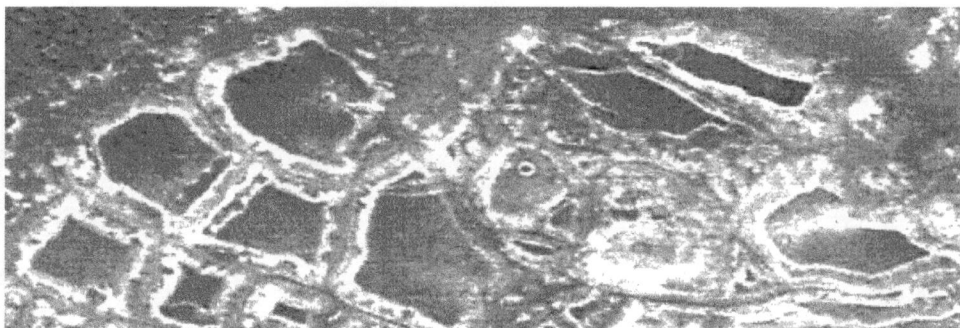

While the reservoirs seem to be dry now, at one time they were filled with water. And the Martian water distribution system operating by way of a series of intricate subterranean pipelines was working at capacity.

From where did all of this water come and where did it go? This will be fully examined after taking a slight cosmic detour.

8 Guardians of Mars and Moons

Some mysterious force has been protecting Mars and its Moons. It isn't known for how long this guardianship has existed but its protection of the red planet and its satellites has been quite obvious.

While the more recent Earth missions to Mars have been extremely successful, this has not always been the case. In fact, it is only until about 2010 that attempts to reach the red planet even managed to reach the red planet. A long succession of failures – many of them of strange occurrence – haunted missions to Mars prior to the latest successes. Not only did these befall the United States, but any other nation which attempted to penetrate Martian space.

It's almost as if a curse – a Martian curse – had been placed on any flight directed at the red planet. The word curse is used with purpose because in many of the failures the culprit seemed to simply be bad luck directed by an intelligent entity. Who is the source of this bad luck? Before examining this, first study the very long list of failures of attempts to explore Mars. And, at the same time, ask yourself: what was the compulsion to keep sending mission after mission toward Mars in the wake of so many disasters?

CHRONICLES OF FAILED MISSIONS TO MARS

Marsnik 1 (USSR) launched Oct. 10, 1960. It was supposed to be a Mars flyby but the spacecraft did not reach Earth orbit.

Marsnik 2 (USSR) launched Oct. 14, 1960. Another intended Mars flyby which did not reach Earth orbit.

Sputnik 22 (USSR) launched Oct. 24, 1962. Another planned Mars flyby where the spacecraft only achieved Earth orbit.

Mars 1 (USSR) launched Nov. 1, 1962. A planned Mars flyby. The spacecraft's radio failed at 65.9 million miles.

Sputnik 24 (USSR) launched Nov. 4, 1962. Intended Mars flyby. Spacecraft achieved Earth orbit only.

Mariner 3 (U.S.) launched Nov. 5, 1964. The shroud encasing the spacecraft at the top of the rocket did not jettison.

Mars 1969A (USSR) launched March 27, 1969. The Mars orbiter did not reach Earth orbit.

Mars 1969B (USSR) launched April 2, 1969. Orbiter failed during launch.

Mariner 8 (U.S.) launched May 8, 1971. Orbiter failed during launch.

Kosmos 419 (USSR) launched May 10, 1971. Mars orbiter only reached Earth orbit.

1971 – Russian Mars 2 orbiter, launched May 19, 1971. Lander crashed. Mars 3, launched on May 28, 1971 and arrived Dec. 3. Briefly worked, then died out.

Mars 4 launched July 21, 1973. The Mars orbiter flew beyond Mars on Feb. 10, 1974.

 Mars 5 launched July 25, 1973. Arrived on Feb. 12, 1974, but lasted only moments.

Mars 6 launched Aug. 5, 1973. Flyby module and lander arrived on March 3. 1974 but the lander failed when it crashed.

Mars 7 launched Aug. 9, 1973. The Mars flyby module and lander arrived on March 3, 1974, but the lander missed the entire planet.

Russian Phobos 1 launched July 7, 1988. The Mars orbiter and Phobos lander were lost.

 Phobos 2 launched July 12, 1988. The Mars orbiter and Phobos lander were lost in March 1989 near Phobos. **Phobos 2 was shot down by a UFO.**

American Mars Observer was launched to the planet on Sept. 25, 1992. It was lost just before it was supposed to achieve orbit on Aug. 21, 1993. Unexplained loss of communications was the problem…it was assumed.

Russia's Mars 96 was launched on Nov. 16, 1996. However, the orbiter, two landers and two penetrators were lost after the rocket failed.

Japan attempted a mission to Mars with Nozomi, which launched on July 4, 1998. The orbiter failed to enter orbit in December 2003.

Mars Climate Orbiter was launched on Dec. 11, 1998 but **vanished** after arriving in September 1999.

Mars Polar Lander (MPL) and two penetrators with it (called Deep Space 2), were launched on Jan. 3, 1999. They too were also lost when an engine turned off prematurely.

Russia sent a Phobos–Grunt mission to Phobos which was launched in 2011. It crashed on January 15, 2012 after not reaching orbit. It was also carrying China's first attempt at a Mars orbiter called Yinghuo–1.

Quite a list of failures! Why do these various space agencies keep trying so desperately to reach Mars? That is a secret that no one from either NASA or the other space agencies have been willing to share.

But there is more to this story of repeated failures. The cause of more than one of them was by direct interference from UFOs. In fact, the end of the PHOBOS 2 mission in 1989 was directly caused by a UFO and it is still a mystery why this has not been touted as absolute proof of the existence of extra–terrestrial life.

In January of 1988 Russia sent two probes to investigate the moon Phobos: Phobos 1 and Phobos 2. Phobos 1 was lost in space while Phobos 2 reached orbit around Mars. However, on March 28, 1989 communication with Phobos 2 failed. According to the Soviet News

Agency Tass: "Phobos 2 had failed to communicate with Earth as scheduled after completing an operation yesterday around the Martian moon Phobos. Scientists at mission control have been unable to establish stable radio contact."

Just before all communication was lost with the Phobos 2 probe, the spacecraft sent back to Earth this final photograph as it circled the Martian moon:

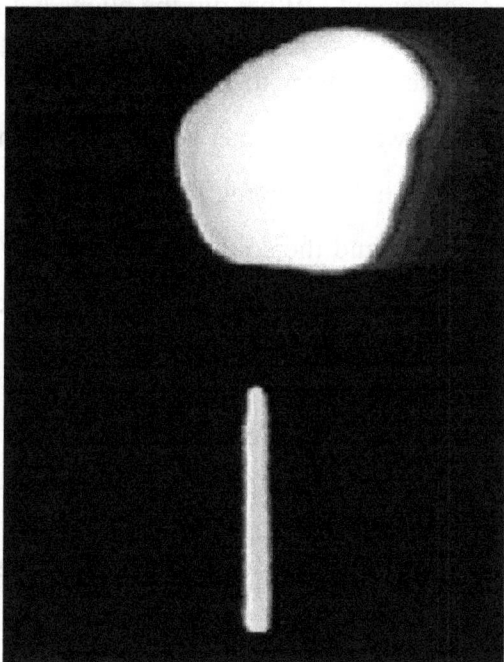

The tube shaped object in the bottom of the frame approached Phobos 2. Moments later, Phobos 2 was gone. Not only had radio communication stopped, but the Russian craft was gone. Not a trace of it has been found or photographed by any of the successful probes to Mars or its two moons.

As of this writing, the United States has its rovers crawling freely over the red planet. The missions to Mars of late have been amazingly successful. There doesn't seem to be an outside force trying to prevent exploration of this region of space any longer. Why not?

It seems obvious that some form of agreement has been arranged between the United States and alien agencies which has allowed unimpaired study of the planet Mars and its two moons. What was the other portion of this agreement? What did the aliens get in return?

And why this desperation to colonize Mars? Could it be that climate change – global warming – has gone beyond the critical stage at which it can be stopped? Is Mars being prepared as the new home for the billionaires, the super–rich and the soulless politicians? And is it possible that once on the red planet these "lucky" people will become the property of the grateful aliens who truly own Mars?

Why has Mars suddenly become available for settlement by people from Earth after all of the previous sabotaged missions?

And what about the mysterious Martian moons? What is their secret?

The moons of Mars – Phobos and Deimos – were officially discovered in August of 1877 by Asaph Hall. Deimos was discovered on August 12th and Phobos was discovered on August 18th. Mr. Hall had been purposefully searching for the moons of Mars after having spotted on August 10th what he thought might have been a Martian moon. But he was prevented from further observation on that or the following night by bad weather.

The original names of the moons were first spelled Phobus and Deimus (fear and freight) as in the Iliad. The discovery of the tiny moons was made at the US Naval Observatory, using a 26–inch refractor telescope.

Since the moons were discovered in 1877 it is difficult to explain how author Jonathan Swift in his book "Gulliver's Travels," which was written in 1726, accurately described the existence of two small moons circling Mars (Part 3, chapter 3 – Voyage to Laputa). Some people claim that he was simply guessing and was following the widely held belief at that time which dictated that Mars should have two moons because that was the logical progression of satellite acquisition of the then known planets. This makes little sense when considering that neither Mercury nor Venus have moons, Earth has one and Jupiter had four at the time. Why would Mars have two?

Another interesting theory was made by well–known Russian spacecraft designer V.G. Perminov which states that Swift discovered documents that had been left on Earth by a Martian describing the two moons.

The brilliant philosopher and writer Voltaire also placed two moons in the skies of Mars in the 18th century in his short work called "Micromegas" in which an alien visitor to the Earth talks about them. Some people say he got the idea from Swift, some from Kepler. Some people think it was just a lucky guess. Was Voltaire using the principal that Mars should have two moons because of the same progression of satellite acquisition theory that Swift may have followed?

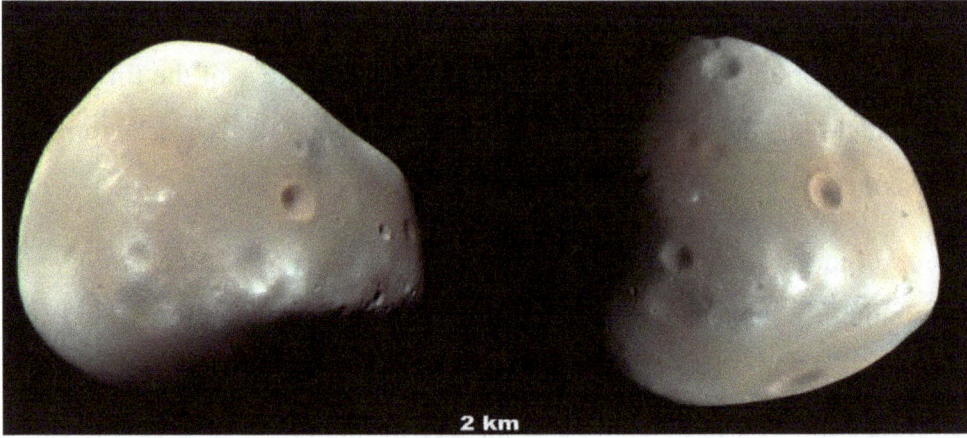

2 km

DEIMOS

In recent years, many people have espoused the theory that both of the Martian moons are artificial. Some have even speculated that they are hollow. Those who believe this are all still being fooled by the trick played upon the world on April 1, 1959 by amateur astronomer Walter Scott Houston. It was in 1959 that he proposed the artificial satellite theory in the publication *Great Plains Observer*. The proposal was made as an April Fool's joke. The hoax spread farther than he had anticipated when renowned Russian scientist Iosif Shklovsky circulated the story as being genuine.

These two moons play a very important role in the theory being proposed in this book. It isn't suggested that they are artificial. But what is considered to be possible is that they were natural objects existing in the asteroid belt that had been maneuvered by alien beings into position around the planet Mars. Why? To stabilize the out–of–control obliquity of the planet in order to provide a more uniform environment for the planned alien colony. These were the most recent inhabitants of the planet Mars as soon to be considered.

When did the aliens place Phobos and Deimos into orbit? It would most likely have occurred after the catastrophe which struck Mars and threw it nearly sideways on its axis.

9 Zone of Peril

Mars has always balanced on the edge of destruction. It is surprising how little consideration has been given to the red planet's precarious position being so close to the asteroid belt. Not only can it be struck by asteroids being randomly jolted out of place by the gravity of Jupiter but many of the planetoids regularly cross the orbit of Mars.

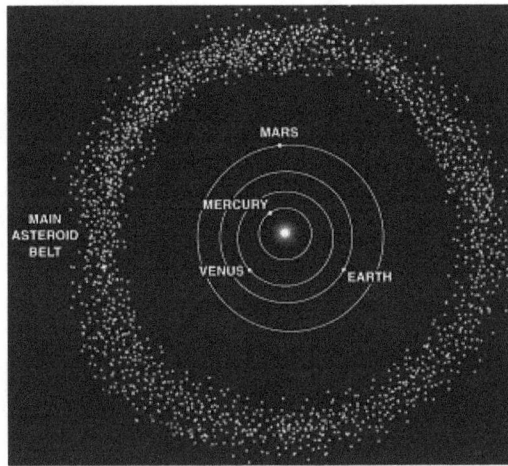

Where did the asteroid belt come from? One belief is that it is the remains of a planet that had met its destruction by some means long ago. It might have been struck by another planetoid. Or it might have been torn apart by the mighty gravitational pull of Jupiter. Some imaginative people even believe that the planet that once existed between Mars and Jupiter was destroyed by some type of war waged by its inhabitants.

There is another theory similar to the last one which states that the supposed planet had actually been inhabited by Satan and his legions and that it was destroyed by the Almighty. The victims were the fallen angels who were either cast down to Earth or took up residence on the nearest planet, that being Mars, where they continued their evil activities.

One of the names given to this annihilated planet was Phaeton – suggested by Russian scientist Yevgeny Krinov who studied the Tunguska Event – and many books have been written about Phaeton and its former demonic inhabitants. While these works make fascinating reading, they lack in scientific evidence.

It is claimed that the asteroid belt could not have been caused by the destruction of a planetary sized body because there currently does not exist enough material in the belt to account for anything larger than our Moon. And almost fifty percent of the mass of the asteroid belt is currently contained its four largest members: Ceres, Vesta, Pallas and Hygiea. The important word is currently.

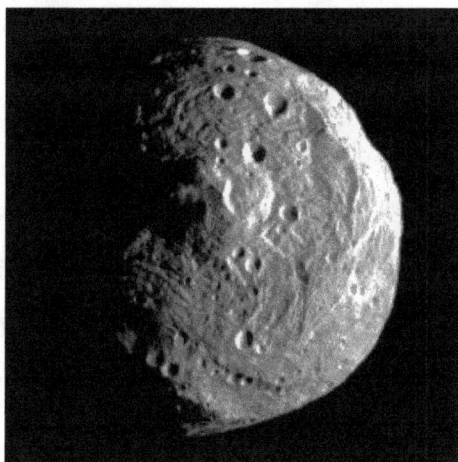

VESTA

A few million years ago the asteroid belt consisted of at least twice the amount of material that it now holds. This mass was gradually lost over the years through the gravitational effect wielded by Jupiter, casting the material into the deeper reaches of space. Thus, eons ago there had been enough mass in the asteroid belt to account for a sizeable planet to have existed there and as such filling the Titius–Bode Law of distance between planets in the solar system.

However, there are a couple of major problems with the belief that the asteroid belt was caused due to the destruction of a planetary body. One problem is that the amount of energy that would have been generated due to such a destructive force being unleashed would have upset the balance of the entire solar system. A great deal more damage would have been produced.

Another major and deciding problem with the theory that a planet was destroyed in this zone is that the chemical composition of the material in the asteroid belt is not uniform. It did not originate from a single planetary body.

It seems clear that the asteroid belt is made up of material from the original creation of the solar system and simply never accreted into a planet–sized object. And thus a constant threat was left just outside the orbit of Mars. What were the odds in favor of one or more of the larger objects in the belt breaking free and plunging one day into the red planet?

10 The Native Martians

Mars could have been inhabited by humanoid like beings for millions of years before primates even existed on Earth. For some reason, we have been argued into thinking that Mars has always been uninhabitable for humanoid like beings. And it seems that most people used to simply accept this without really thinking it through.

Consider the Martian atmosphere. Mostly carbon dioxide. Not pleasing for Earthly humans, not very rich in amount, but why couldn't a native Martian have adapted to it? We are told that the air pressure at the surface is much too low for anyone to survive in such conditions. We are told that it is equivalent to what exists at the heights in the Andes Mountains. Why are we not at the same time reminded that people like the Incas of Peru lived quite comfortably at a similar altitude? What would prevent a Martian on Mars from doing the same on his own native world?

And we're convincingly told that it was too cold for any human like being to survive on Mars. Too cold! Right now the temperature can reach about 72 degrees Fahrenheit at certain locations on Mars during the day. While it may drop to 200 below zero at night, the resourceful Martian probably would have found some type of shelter. Don't you think? After all, don't people living in Nepal have to endure some pretty extreme temperatures too? So did the Neandertals on Earth as well as modern day Eskimos.

But, returning to the concept of ancient Mars, isn't it likely that any species native to this world would have gradually and steadily adapted to its continually changing climate? Why not? In fact, many of the sites discovered there are Incan in type. Is that not what we would expect?

The main question should be: is there any physical evidence that Martians had at one time inhabited this planet? Already revealed are structures like the pyramids which clearly show that at one time a society vaguely similar to ancient Egypt was active on this planet. These have already been examined in this respect.

Among the more recent apparent discoveries is at least one example of cyclopean statuary that also was common to the Egyptian world. Note the photo below taken of an intriguing sculpture supposedly found in the side of a Martian mountain by one of the rovers.

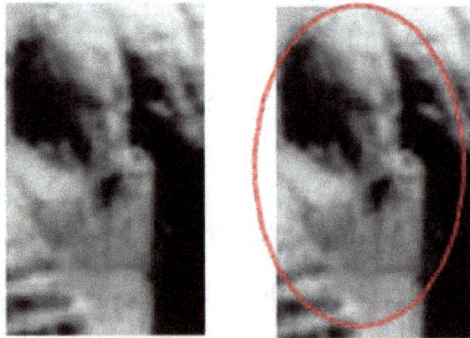

In Ancient Egypt

A closer comparison:

On Mars

On Earth

There is a problem with the Martian version. It seems too good to be true. Firstly, the context is non–existent. In other words, the sculpture appears by itself in the middle of barren wasteland. Secondly, it is too precise a match for the Egyptian counterpart. While independent invention explains that different cultures can produce very similar artifacts it does not suggest that the artifacts are exactly the same. For example, the pyramids of Mexico are like the pyramids of Egypt in form but not exact copies.

There are two reasons why the questionable Martian version of cyclopean statuary was introduced here. One reason is to imply a similarity of environment – being desert type – which would be more likely to inspire similar artifacts. And secondly to suggest the possibility that the civilizations of both early Mars and of ancient Egypt were influenced by the same extra–terrestrial visitors. But, from a scientifically inspired archaeologist's viewpoint, the answer is probably still a matter of independent invention. And again, this simply means that both cultures – the Egyptian and the Martians at this particular time and location – developed along the same path and as such developed similar but not exactly the same relics.

Egyptian obelisk

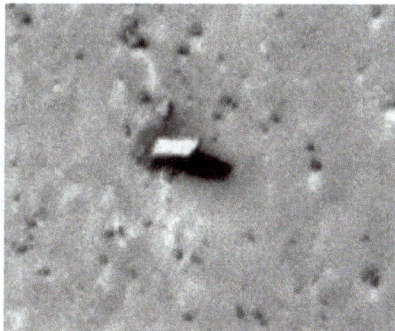

Martian obelisk?

The Martian obelisk may be genuine but not of the same type as that found in Egypt. Another intriguing find on Mars within the context of other similar relics was what appears to be a sculpted head.

What is striking about the configuration of this head is that it is similar in feature to the famous face. They could be relatives. This would add to the likelihood of authenticity. It is consistent. There is context. But, in another sense, the figure seems almost too good to be true. Also, the hairline at the back seems a trifle questionable. I would rate this as probably not genuine.

As in ancient Egypt, the ancient Martians continued technologically advancing. Based on the remaining examples of their culture it appears that on Mars civilization reached a mid–1950's pre Space Age type of level before being annihilated by a fateful planetoid impact.

Below are some examples of what appear to be the remains of industrial objects which were found in proximity of one another, adding to the context equation.

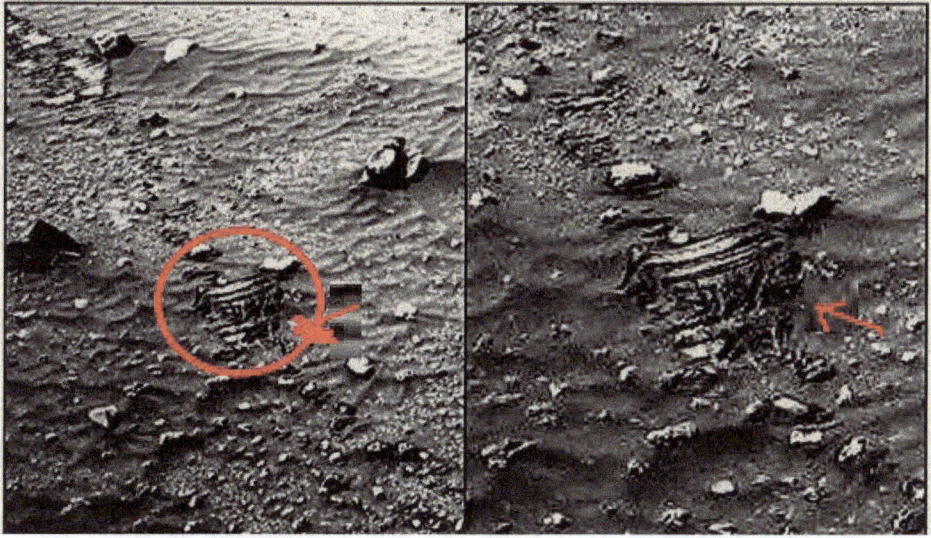

Below is the Y valve again and accompanying features, found in similar
context:

Metal of some type jutting outward from a lid:

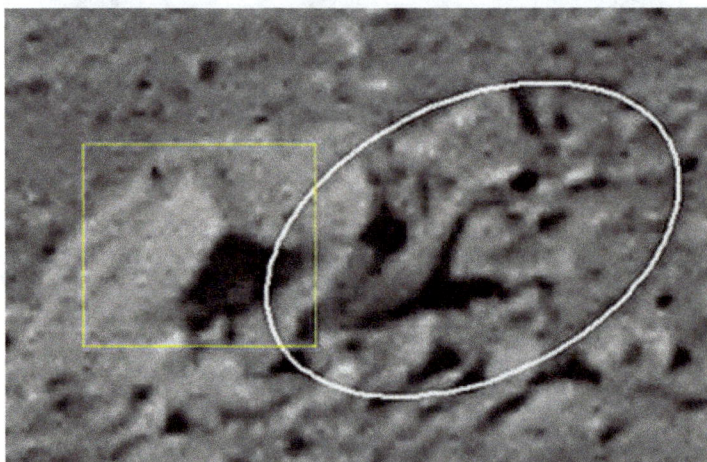

It appears, that on Mars, there remain leftovers of societies which had become stalled in a more ancient Egyptian like status. This can be seen in the remains of one of their towns which barely escaped obliteration due to the planetoid strike. In the picture below this city can be seen precariously on the edge of the great Valles Marineris gorge.

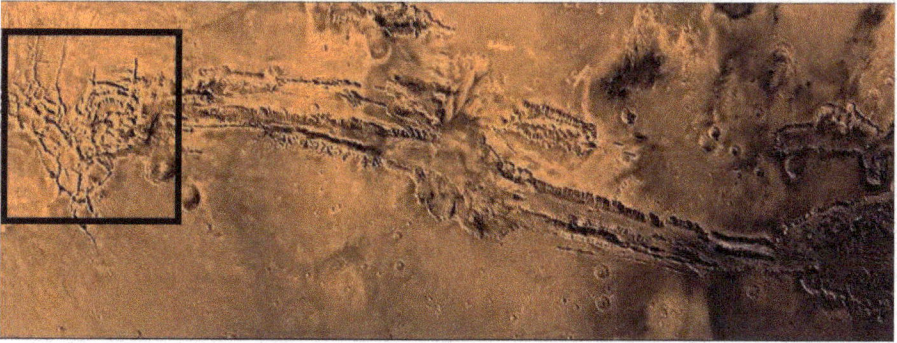

A closer view:

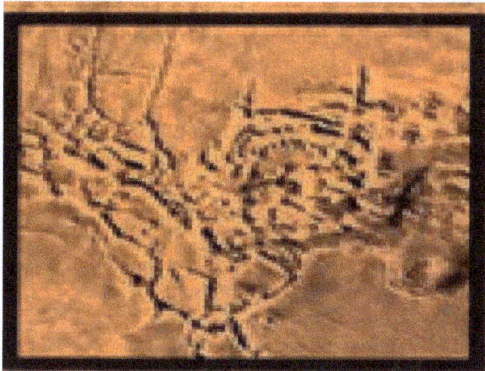

As an archaeologist, I would compare the above view of the potential Martian city with the view below of a genuine Incan city.

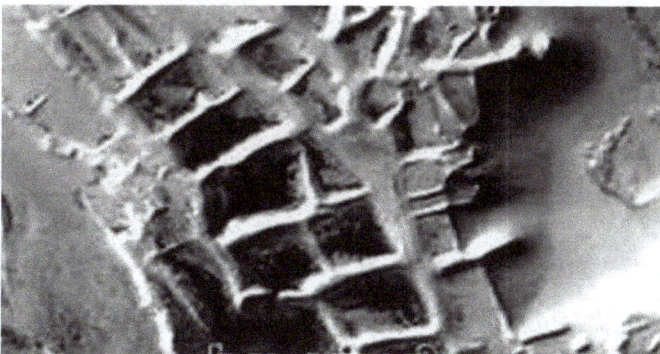

Or the one on Mars in X–Ray:

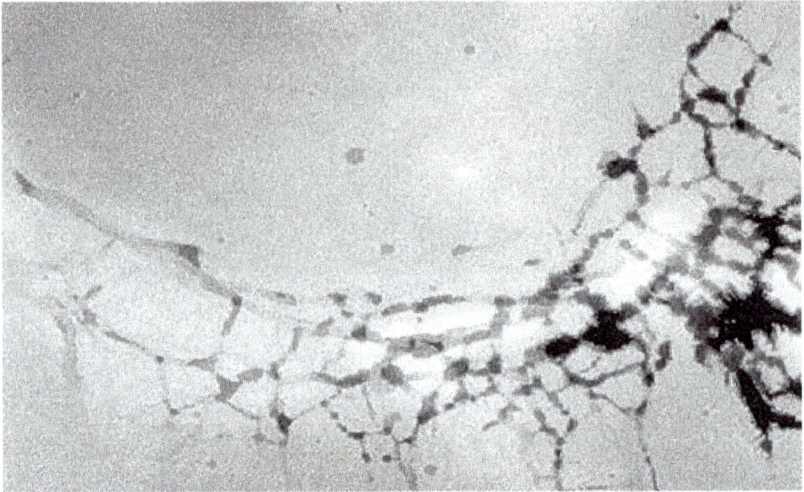

Keep in mind that the air pressure on the surface of Mars is supposed to be similar to that in the heights of the Andes where the Incans resided. Maybe this factor would cause these two societies to construct similar dwellings? Parallel development.

Another potential Martian town from orbit:

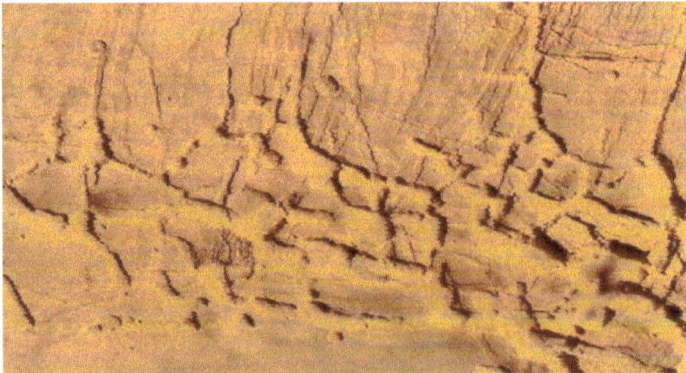

As an example of a town advancing toward the Industrial Age level, the next location reveals a "factory" type of area found in the

Nylosyrtis Region of Mars. It appears to clearly be populated by a number of commercial sized kilns (blackened areas).

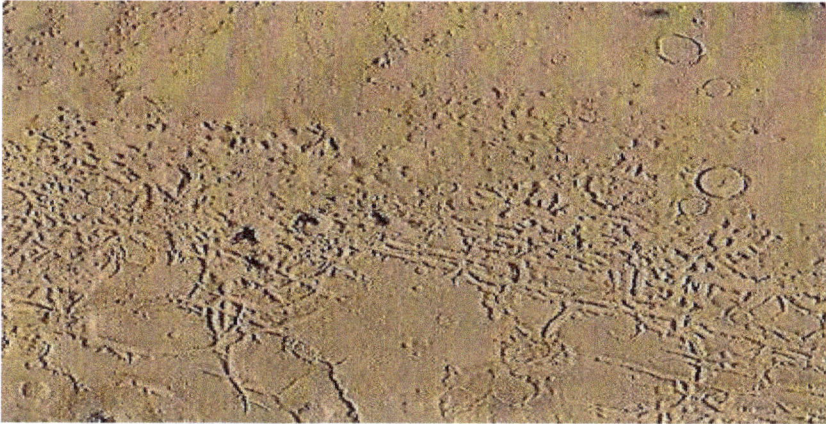

Why are kilns so important to this study of Mars? Simply because these are one of the primary features located among the ruins. And it seems logical. Kilns were used for the creation of all type of ceramic articles and would be a common manufacturing tool to be part of a developing culture. Below are a couple of illustrations of Earth based kilns.

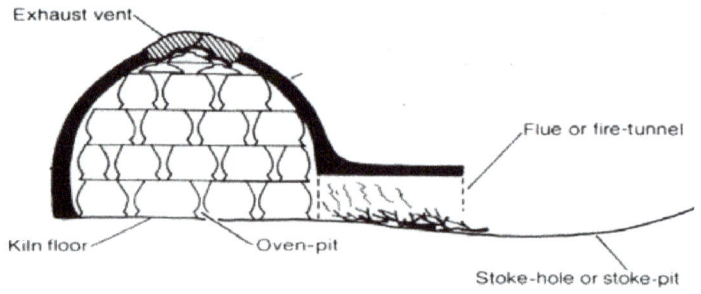

Exhaust vent

Flue or fire-tunnel

Kiln floor

Oven-pit

Stoke-hole or stoke-pit

A close up of what may be a kiln site on Mars.

Once again, the argument becomes even more convincing when comparing an archaeological site on Mars with one of a similar nature on Earth. Below are two photographs of areas of heavy kiln use that occurred in Europe during the late Medieval Period.

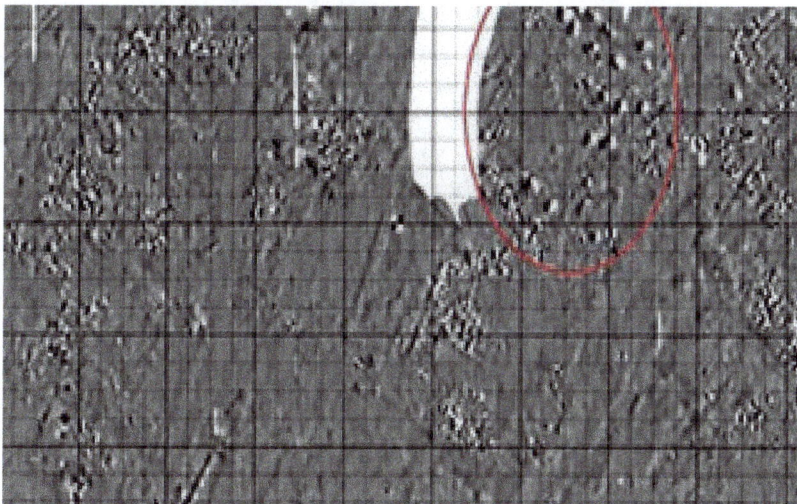

Kiln, or Medieval factory site, on Earth

Kiln – early factory site – on Earth.

Enlargement of an accompanying city which was home to the workers at the kiln site.

Different view of village around kiln location.

A wider view of the Nylosyrtis Region where kilns may have predominated on Mars:

Keep in mind that the remains of the above type of town are not the same type as the remains from the Incan type of village seen earlier. The city above was from a more advanced period in Martian history and has features that are unique to this environment and not necessarily a duplicate form of any Earth city of a comparable time period.

Much of what remains of the original Martian civilization are bits and pieces scattered across the rocky, dusty landscape. They had been preserved over the eons by the layers of sand that had been poured upon them by the wind. And it is the same wind that has revealed them to the cameras in space and on the ground today.

Even though Mars has a type of climate it is not the type that erodes into oblivion evidence of past civilizations. Mars is a dry, dusty world which is blasted by horrific winds. Over the eons the winds have buried many of the archaeological sites and as such preserved them. And these same winds also periodically strip away the covering of dirt and soil to reveal the sites again. This is how the remains of a civilization that may have been destroyed millions of years ago can still exist and be photographed for study.

Martian dust devil

Next to be reviewed are harbor like images from ancient Mars. But first, for comparison, will be seen harbor views on Earth taken from space.

Note the shape of entire area.

Again, note the shape of the area.

The following images were taken from Martian orbit decades ago. They have remained unnoticed by most people because most people are not archaeologists and might not be aware of their significance. But to an archaeologist who is comfortable with studying aerial images for the existence of potential relics they reveal a sometimes startling site.

This is the Ganges Region on Mars where a photograph from orbit shows what appears to once have been a massive harbor area with accompanying small cities.

Below is a close up of what appears to be a small town that once bordered this massive harbor.

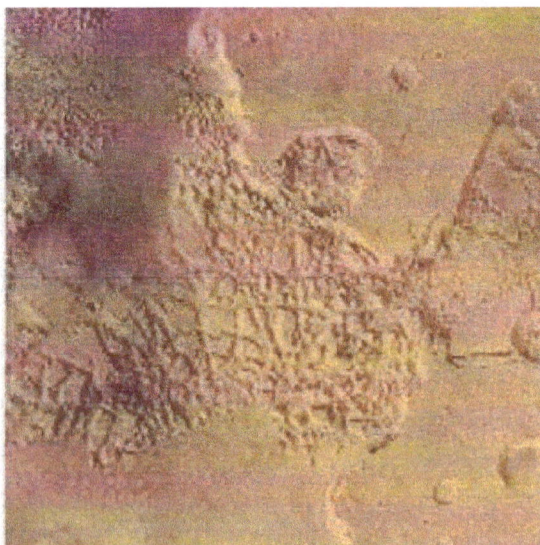

Enlarged view:

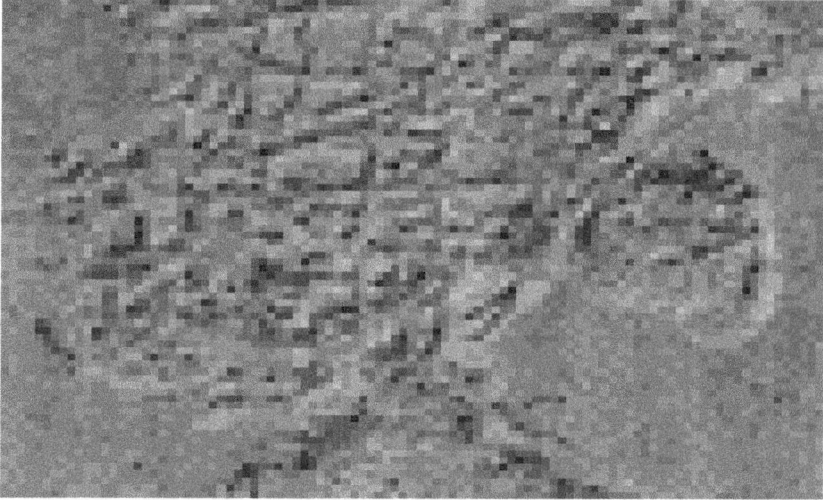

This may appear to be nothing more than a jumble of mounds from a dried out former river bed but they are more than that. Notice the regularity of the structures and the angular forms. It would not have been an Incan type of village, nor a more modern one, but similar to an ancient Roman or Greek city. Below is an example of what the above city may look like from ground level.

Ancient harbor city on Cyprus.

Below is a modern photograph of a formerly deluged harbor town on Earth: compare with the Martian remains.

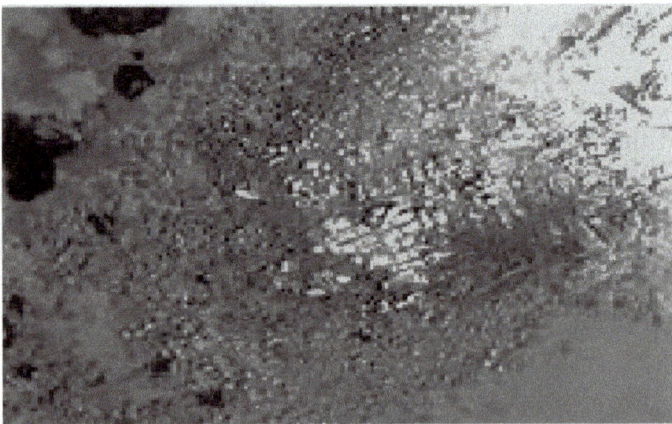

Earth

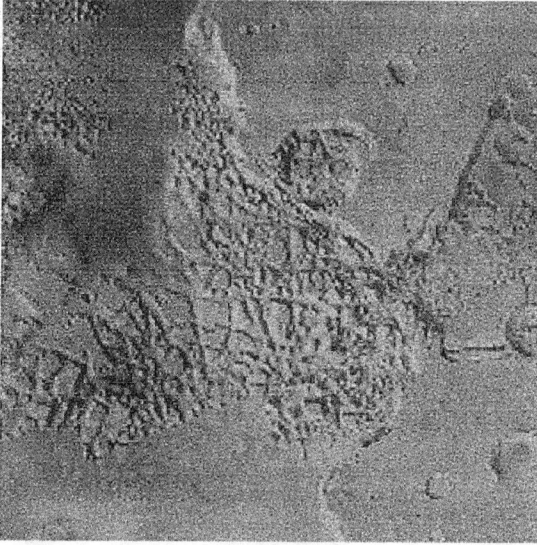

Mars

Examining the harbor area itself it is clear that along the edges of what had been a deep body of water are what appear to be the remains of buildings or dock like structures. Keep in mind that after more than a million years of blasting by the high powered Martian winds the sides of this bay would have been smoothed down rather than being jagged edged.

Ganges Bay on Mars

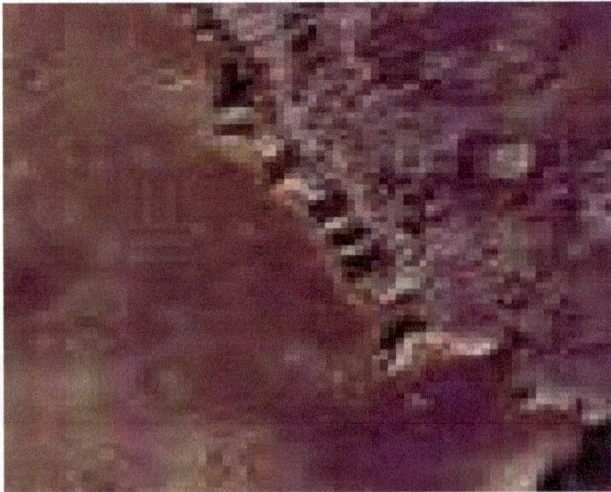

On the edge of the bay

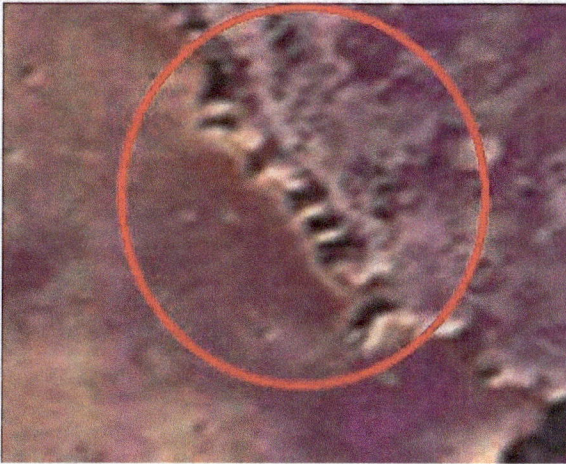

Buildings? Docks? Ships? A combination of all of these? Two massive supports at the end of where a bridge used to span?

Take special note of this bridge feature displayed below. It appears to be both ends of a former bridge abutment. Both sides of the abutment survived. Or can it be simply coincidence that there are two matching pieces of cliff this strangely shape!

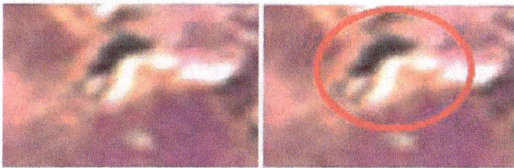

One side blackened, the other white.

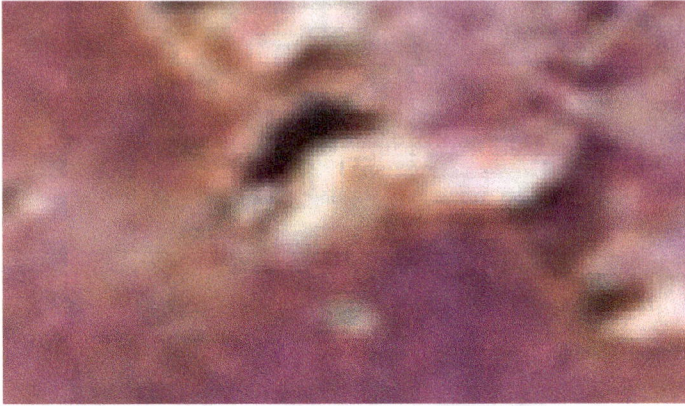

It is difficult to consider this as being a natural feature combined with all of the other non–natural features in the same location.

The ultimate question to be asked is how could it be coincidence that the exact feature that would be expected to be located in this location was in fact observed there? The image of the abutment of a bridge is captured in a photo at the edge of a bay where a bridge would be expected to be. Not just part of the support system but the pair of them side by side! This even negates the idea that these might be natural features. Side by side and exactly the same in appearance except for color!

Wider view of bay port area.

Before leaving the subject of buried ancient harbors of both worlds, notice the next picture of one that was found in ancient Greece:

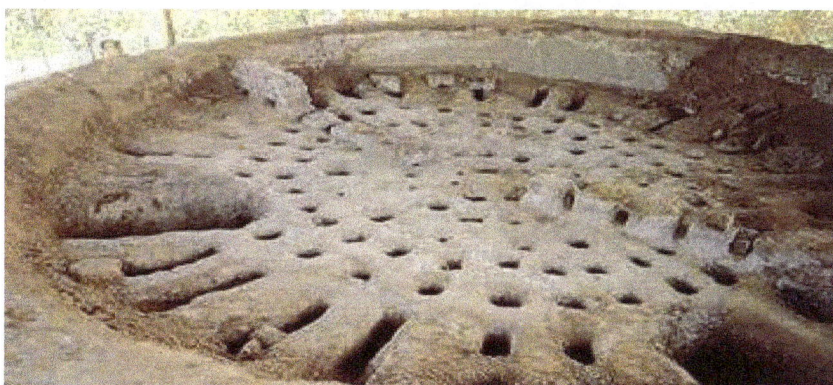

Does not the area on Mars shown next seem similar?

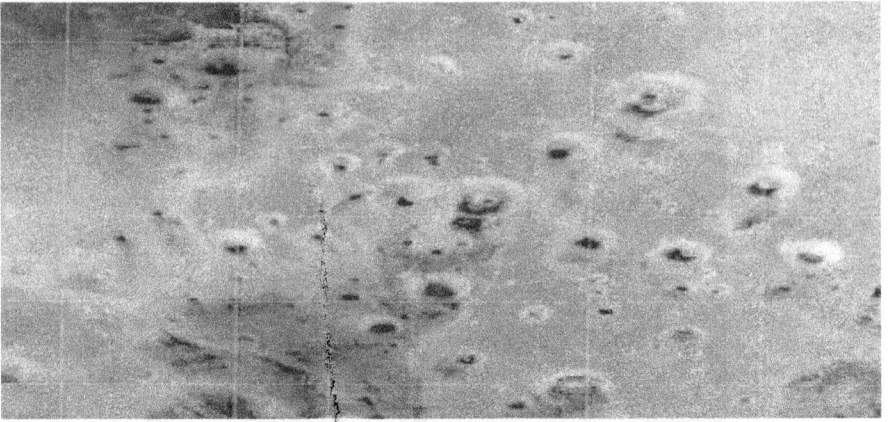

It seems clear that Mars was once inhabited. But what happened to these original Martians? What great cataclysm destroyed their world?

11 Chaos on Mars

About 65, 000,000 years ago

Astronomers had been warning of the approaching asteroid for months. A planetoid–sized mass of stone and iron had become dislodged from the asteroid belt between Mars and Jupiter. There had been many other asteroids ejected toward the red planet before but none nearly as large as this one or at such a deadly angle of impact.

Now it was visible in the sky and there was not any means of escape. Space flight had not been developed on this Industrial Age world and the closest habitable planet – Earth – was 55 million miles away. Neither Phobos nor Deimos were options for escape because they were not even in orbit yet.

Make shift underground shelters were available across the planet for the 2,500,000 inhabitants of Mars and had been often used in the past for less destructive strikes from the sky. But the planetoid bearing down on Mars this time was so huge that it could cause the entire upper crust of the plane to collapse, making the shelters into graves instead of places of refuge.

The monstrous asteroid struck in midday. The entire planet rocked. The impact was so mighty that Mars was shoved sideways on its axis. It was turned almost completely on its side, with north facing where west used to be. This accounted for the new highly eccentric obliquity.

Perhaps it was a heart–shaped asteroid like Itokawa above which impacted.

A rift in the ground was torn where the asteroid struck and would later widen and deepen to form the vast Valles Marineris (below).

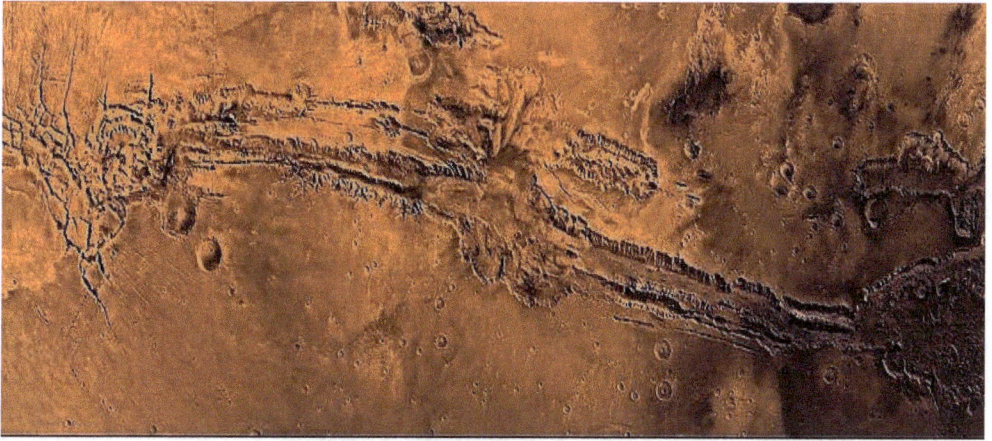

Slightly north of here the three fearsome volcanoes of the Tharsis Montes Region came into existence, born by the internal seething and churning created when the gigantic asteroid plunged into the interior of the planet. Mars then began to destroy itself from the inside.

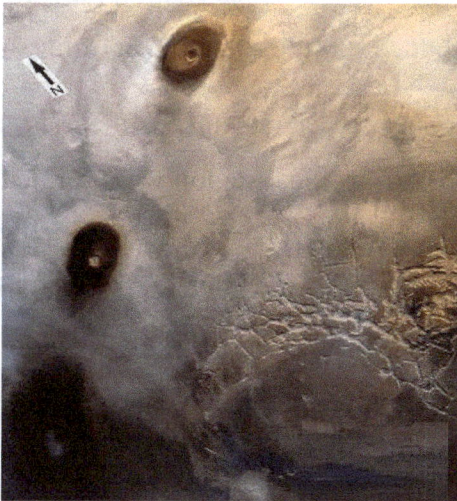

Tharsis Montes

The sun was hidden and the sky was blackened by the millions of tons of dirt and rock that had been thrown into the heavens. Newly enlivened volcanoes poured more soot and smoke into the air and earthquakes tore the landscape apart. The sudden change in the tilt of the planet tossed its seas into space and ejected the atmosphere as well. Devastation was rapid.

Very few people survived this onslaught of nature and those who did found an inhospitable world which was in the midst of its own death throes. It was a strange, hostile world on which they could no longer exist. Such perished the native inhabitants of Mars. All that remained of them were some of the relics of their civilization, the rest of which having been destroyed by the worldwide upheaval.

The preceding account was fiction. However, it was based on a realistic possibility. Valles Marineris may have originated from an asteroid strike and the Tharsis Montes Region may have developed due to the internal chaos produced by the same impact from space.

There are many other scientific theories about how Valles Marineris and the Tharsis Montes Region were formed but they too are all based on speculation. In fact, the "scientific" theories change frequently, based on the prevailing mood of the day. In reality, the theory just presented in this book is probably more likely than the others because it is known that Mars was and still is subject to asteroid impacts and such impacts could cause the type of planet wide destruction as described.

It is as yet unconfirmed whether or not an existing civilization was spread across the planet at the time of the proposed asteroid impact. However, there is evidence in the form of physical remains already revealed that tend to prove that there was.

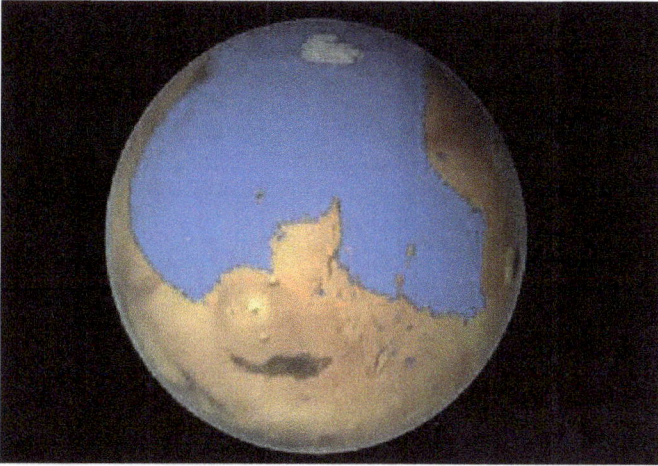

Before the catastrophe

After the catastrophe

It is speculated that this happened about 65,000,000 years ago at the same time that the dinosaurs were being driven into extinction by an asteroid strike on Earth. The reason for this is because it is believed that at that time some foreign object of massive size entered our solar system and upon passing near the asteroid belt forced out of orbit a swarm of planetoids and asteroids many of which bombarded the inner planets. Thus, while the mammals were just beginning to rise on Earth due to the extinction of the dinosaurs the civilization of Mars was coming to a tragic end.

But Mars seemed to harbor life yet again after this disaster. This time brought to the red planet from another world, maybe from another galaxy.

12   Aliens Colonize Mars

Mars was still a very attractive world for colonization to aliens from another solar system. It was an uninhabited planet of moderate size with a climate that was not too extreme. It is fourth in distance among a group of planets from an average star in a stable solar system. About 55 million miles away is an inhabited planet which can be used for a variety of resources if what is needed cannot be found on Mars. This might be how an alien species from another solar system would have viewed the prospects of settling the red planet.

There was at least one major problem with Mars, however. It was almost lying on its side on its axis which meant a highly unstable obliquity. In an attempt to solve this matter the outsider extra–terrestrials may have maneuvered a couple of the larger planetoids from the asteroid belt into orbit around Mars. The intention would be to stabilize the planet by the gravitational counterweights of the orbiting asteroids. It apparently had little effect if this plan was indeed carried out. But the extra–terrestrials remained in control of Mars anyway.

The assumed beginnings of the extra–terrestrial presence on Mars is based on the sudden appearance of Phobos and Deimos in orbit around the red planet. This would place the date at about the start of the 1700's since there were not any previous credible reports of moons around Mars and the first mention of the two satellites was by Jonathan Swift in 1726.

It may even be more likely that the outsider extra–terrestrials used the two moons for other purposes than to stabilize the planet's obliquity. Recalling the many sabotaged Earth–based missions to the red planet, it would have been easier to accomplish such strategy from orbiting platforms such as Phobos and Deimos. Electronic signals could be relayed from either or both of them, to receiving stations on our Moon, then back to Earth or aimed at probes that were on route to Mars.

Both the sabotaging of missions to Mars and the placement of the two moons in the sky are evidence of extra–terrestrial's presence on the red planet. Even should the moons be proved to have arrived in orbit by natural means, they still could have been used as platforms by outsider aliens to sabotage any attempts at colonization by any other species.

Unfortunately, at this time there is only scant physical evidence of an advanced alien race having been on or currently present on the red planet. As already noted, most of the remnants of civilization on Mars were probably the products of the original native population.

However, one form of physical evidence pointing toward alien presence are the subterranean networks as photographed by the MOLA orbiting craft.

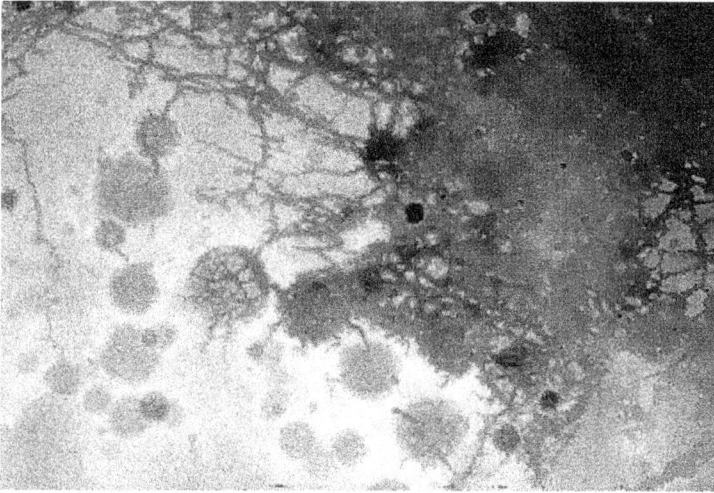

Could this be an underground network of connecting passages leading to various environmental centers? Because it is below ground any alien activity would go unnoticed.

Percival Lowell was able to see only portions of this subterranean network during his observations, and these he probably confused with above ground canals. He was correct in stating that there was life on Mars. However, he may not have been correct about its identity. Instead of native Martian life forms, he was viewing the evidence of alien colonists of the red planet.

But there is another astounding discovery that Mr. Lowell made during his years of intensive observation of Mars. He repeatedly noted in his journals that were written down at the time of observation inexplicable flashes of brilliant light occurring all across the Martian landscape. On one occasion he reported viewing them through the clouds on Mars, a feat of extreme observing that many thought an impossibility with only a telescope from Earth. These findings are of such immense importance that

Lowell's exact account is reproduced below from his personal journal.

On the 17th it was still there, λ 265°, as soon as the planet was looked at. On the 18th I saw it "through cloud." On the 19th it was comparable with Hellas, then conspicuously white, and seemed to be the brighter on the whole. At one moment, λ 281°, a flashing dot shone forth in the centre of it. It was last seen on the 20th, after which the rotation of the planet made it come on the terminator too late.

On August 31 it was still there, and continued so for as much of the September presentation as I was able to observe.

Following is a table of the times the phenomenon was witnessed.

	h.	m.	λ °	
June 16	6	18	289	" Aeria coastline near Cape of Oenotria bright, saliently so."
		37	294	" Aeria near cape white."
17	5	42	271	" Bright spot on Aerial coastline visible."
18	6	48	277	" Bright spot in Aeria visible through cloud; suddenly conspicuous."
19	6	48	267	" Bright spot in Aeria very white, near terminator."
		52	268	" More conspicuously white than Hellas."
	7	0±	270	" Patch of white covers nearly all of Aeria visible."

Phenomena of which this is one of the earliest examples are as difficult of interpretation as they are interesting to observe. They do not lighten with obliquity; they persist for months without diurnal change; and they appear always in definite localities.

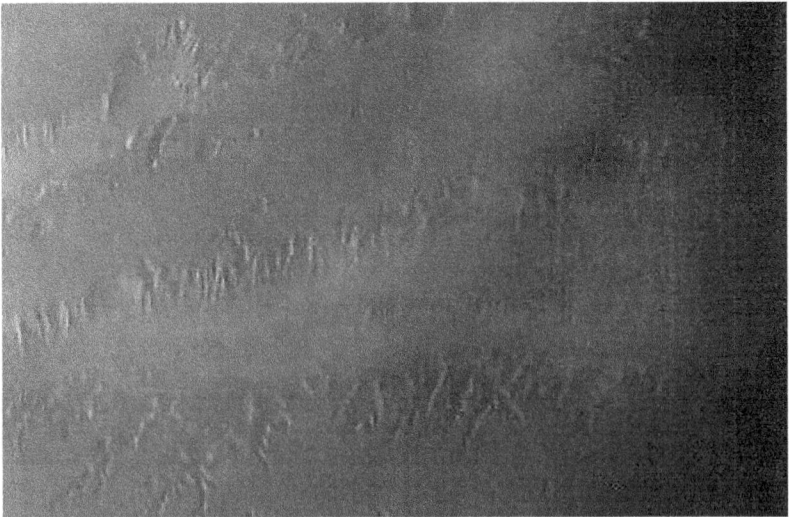

Clouds in Tharsis.

Clouds in canyons

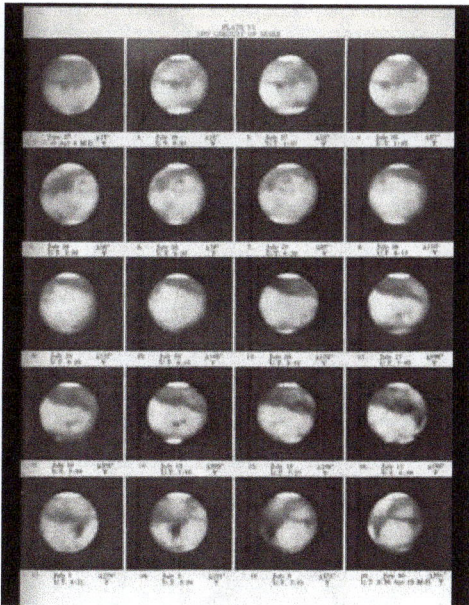
Lowell's original photographs of Mars.

Percival Lowell saw a persistent series of flashing lights on the surface of Mars during his observations of the red planet between 1897 and 1901. And as in the other cases where his sightings were verified, so too were these. The photograph below was taken by one of the Mars rovers more than one hundred years after Lowell's initial report.

Was it a signal like this that was transmitted to an orbiting alien craft which ordered it to shoot down a Russian probe that was heading toward the moon Phobos? Or was it a signal like this that was transmitted to a receiving station on one of the Martian moons which transmitted the message to a similar station on our Moon which then sabotaged the various craft from Earth that were on their way to explore the red planet?

And what more genuine proof can there be that some form of alien life is present on the red planet than the fact that one of their craft destroyed on video an exploratory probe sent from Earth! (See photo). The alien craft is the cigar–shaped object.

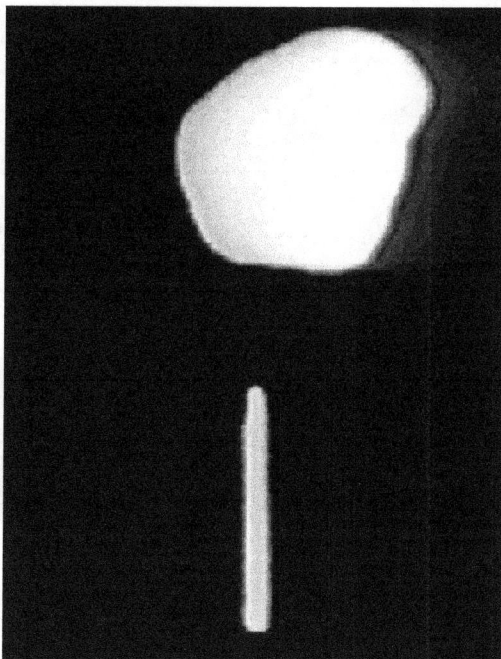

Ironically, it is exactly this type of alien spacecraft that provides the link between the extra–terrestrials on Mars with those who are occupying our Moon. This type of tubular shaped craft has been observed in numerous locations on the lunar surface.

I possess an immense catalogue of lunar photographs, beginning with the original Ranger pictures. Because of this, I also have access to photographs that had yet to be censored by NASA either because they made a mistake or simply were not aware of what was in the photograph. However, among the pictures in this catalogue is one which reveals the existence of one of the alien tubular shaped craft which twenty years later NASA finally decided to censor by placing a black bar over it. But the bar was placed in the wrong spot. Note the UFO below on the crater's edge:

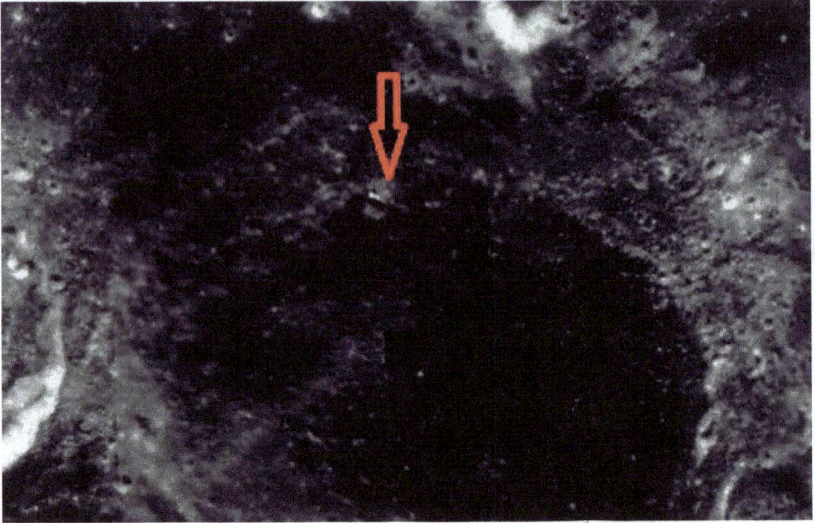

UFO on rim

The top part of the crater rim should have been blackened out where the tubular shaped craft was located but instead the middle part of the crater and the bottom rim were censored.

Tubular UFO on top rim

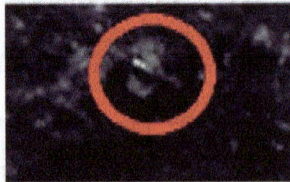

Another vehicle is also pictured in the center of the crater; NASA did get this one correct later. But in an earlier photograph, this other craft is also visible before being censored.

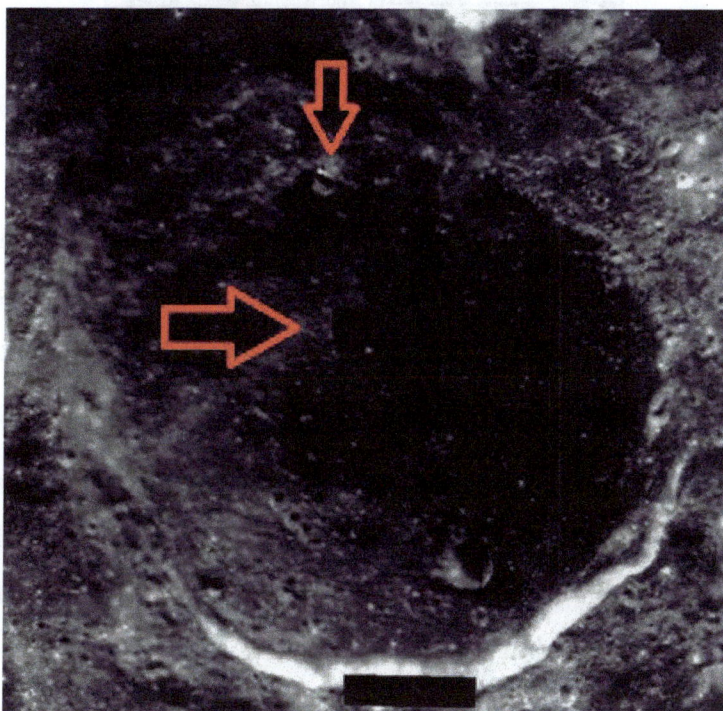

Censored middle of crater and lower rim

There is a UFO in the photograph where the arrow in mid crater is pointing. NASA could not reach back into time and censor the original copy that is in my possession which shows:

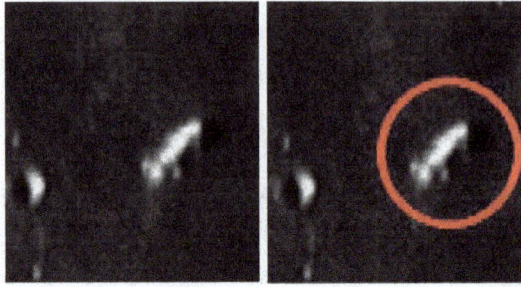

In the decades old uncensored photograph

This is a craft that stands on four legs when needed but has the ability of motion, having been found in several places across the surface of the Moon. The full description of this and other NASA deceptions are found in my other book, "Alien Threat from the Moon."

It has been suggested that at times signals have been transmitted from Mars to Phobos and then to the Earth's Moon. One way this might be accomplished is by using as transmitters structures that look like obelisks or monoliths but whose actual purpose is to deliver electronic, magnetic, or a completely unknown form of transmission between one another. Take special note of the lengthy shadows in the first two pictures below.

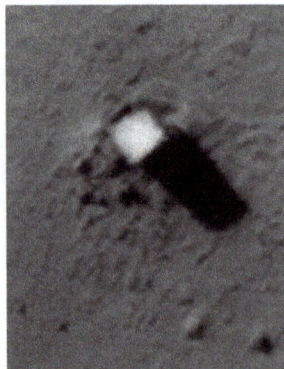

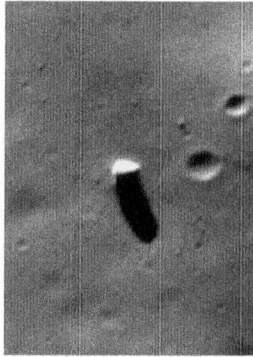

Obelisk on Phobos

Obelisk on Earth's Moon

Is it mere coincidence that these three locations all possess at least one obelisk type structure on their surfaces? Admittedly, this formation is not unknown to occur naturally, but the fact of their existence on each world cannot be discounted as unimportant.

13 Martian Evolution

Based on geological and archaeological evidence, what follows is what appears to be the most logical progression of the rise and the destruction of civilization on the red planet. Inasmuch as Mars formed about five hundred million years before the Earth due to its location in the solar system it can be assumed that any civilization that developed on the red planet did so in advance of civilization on our world. However, instead of developing a much more advanced society it appears that because of a natural catastrophe the evolution of life on Mars was cut short and did not reach a level much beyond that of a 1950's, pre Space Age Earth style type technology. This conclusion is based on the artifacts discovered on the red planet to date.

It doesn't appear that Space Age technology was ever attained by the native inhabitants of Mars which would have allowed them to escape the disaster which struck their planet. Nor would Martian civilization have been advanced enough so as to be able to construct adequate subterranean living facilities in order to survive underground. Unless, of course, Mars is substantially hollow.

The early Martian environment was much like that of Earth's. There was water, a breathable atmosphere which allowed life and civilization to develop in parallel fashion to that on Earth. This is assumed because of the known geological history of Mars which would have existed in a similar climate as that which prevailed on Earth many millions of years ago. This would imply that a similar, but not necessarily identical, species of thinking being would develop along a similar path as had the hominids on Earth. Unfortunately, most of the evidence of this would have been destroyed when the red planet was struck by at least one planetoid from the asteroid belt, shoving Mars sideways on it axis.

The Martian archaeological remains that have been investigated in this book include Incan like cities, Egyptian age habitation zones, mid–European Renaissance period type kiln based manufacturing locations and sites which revealed objects from a more modern 1950s era industrial society. Why so wide an assortment and how did any of the ruins survive to this day?

On modern day Earth different levels of society co–exist. There are still less technologically advanced "tribes" living in the South American rain forests and the pyramids and Sphinx still exist as relics from extinct civilizations. Both of these occur alongside a technologically advanced world which is reaching farther and farther into space. This same situation was true on Mars.

As noted earlier in the book, the particularly severe winds of the red planet are capable of burying entire cities and then scooping them out again a millennium later. This is a perfect way for a site to be preserved: by burial in a dry, cold, barren environment. In fact, much of the surface is often covered by what could be termed dry ice, a perfect preservative.

All of these Martian features have been investigated like a typical archaeological excavation as if they were on Earth with the same results. Some of the sites seem to be obvious remains of civilized habitations;

others were more speculation. But the evidence provided by many of the photographs is too impressive to deny. There was civilization on this world millions of years ago.

But many of the phenomena occurring on the red planet today are the result of a different type of presence, an extra–terrestrial presence. Mars is important to one or more species from another world. Why is not clear. But that they do want to protect this planet from other outsiders is displayed in the many attempts made to keep Earth vehicles from reaching Mars. Why recent missions to Mars have been allowed to progress is also unclear; maybe the aliens are simply setting a trap for us, maybe some form of secret treaty or agreement has been made with Earth governments. For one reason or another rovers are busily scanning the surface of the red planet from the ground while orbiting craft continue their detailed mapping.

Governments of Earth are planning their first "manned" trips to Mars. Is it merely for exploration but is the quest based on fear that humankind may soon be requiring a new home on which to live due to either climatological catastrophe, biological disaster, or thermo–nuclear self–destruction? Either way, we will be the aliens on this red planet. But the memory of the Martians who were there before us remain.

14 Aliens, Masons and Sasquatch

One day before this book was to be published, NASA suddenly released thousands of new high quality photos taken from orbit over the planet Mars. I began scouring and studying them the moment I received them, fearing that anything anomalous would be quickly censored. It turned out – I was correct.

Among these images were discovered many of surprising and some of shocking nature which seemed to confirm the idea that ancient Martians once occupied the red panet. They also strengthed the belief that extra–terrestrials were now in control here.

What follows will be a review of the major new findings made by this author. In a couple of cases, NASA has already censored the photographs that I examined only the night before. These will be noted by using NASA's own script. In it, NASA claims there are no images available for that location, yet the image that you will be viewing is one of those non–existent images!

The first location to be viewed is Richardson Crater. As you will note, there appears to be an exceptionally long pipeline of some type in this location.

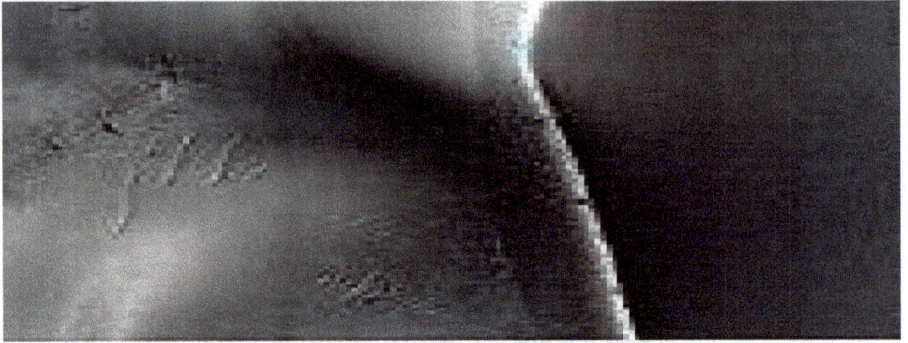

Richardson crater

Take particular note of how the comparison between the pipeline on Mars and the one on Earth both have the same type of connecting and stabilizing features as pointed out by the red arrows.

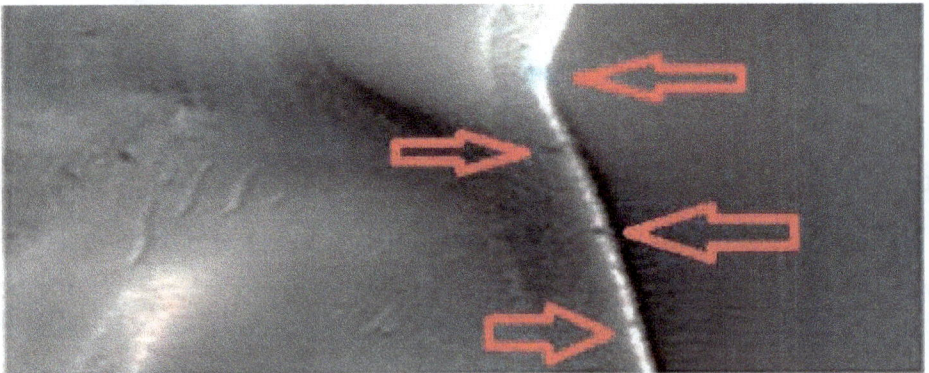

Richardson crater

Pipeline on Earth

There was another pipeline to the left of the one above but it was either never completed or was destroyed in an accident.

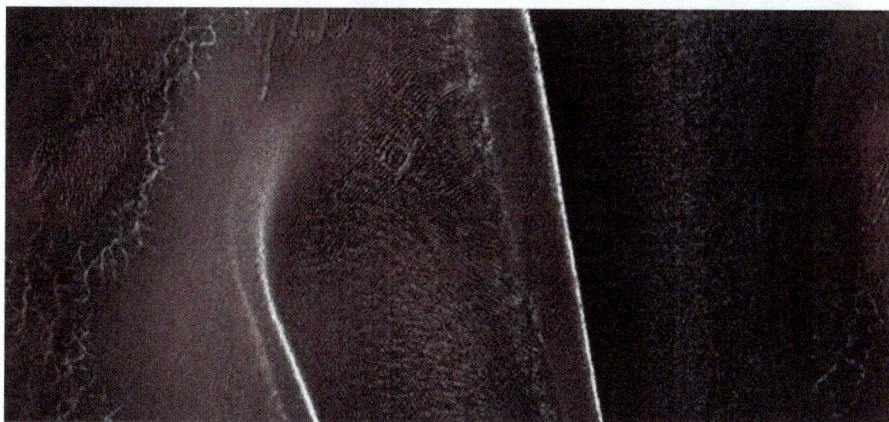

There also appears to be what looks like a former city of rather large size between the great pipelines.

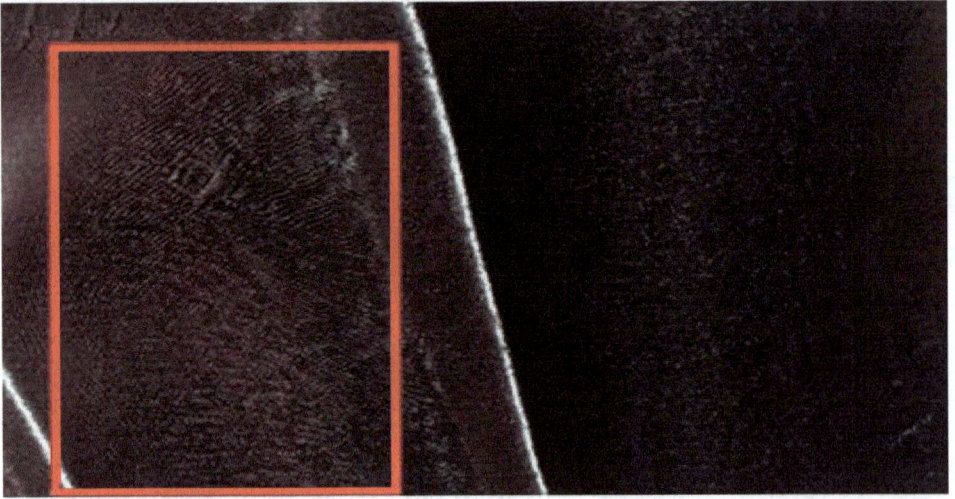

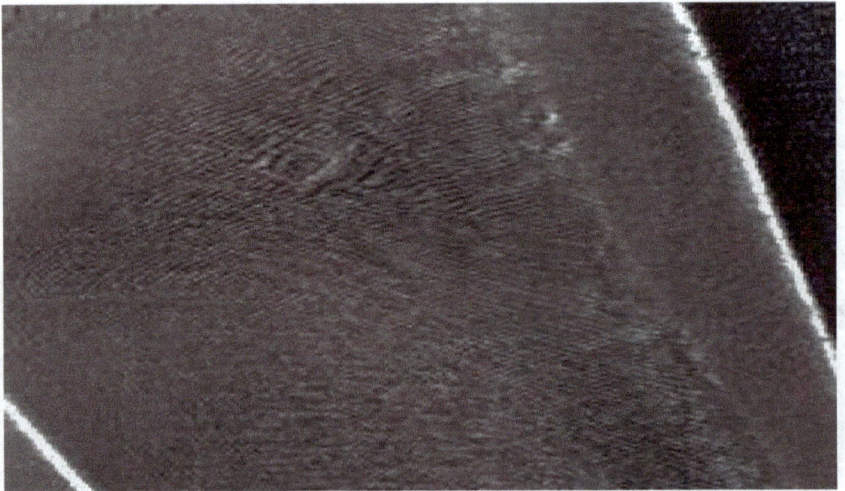

City?

Below is a close up of what I term the city center.

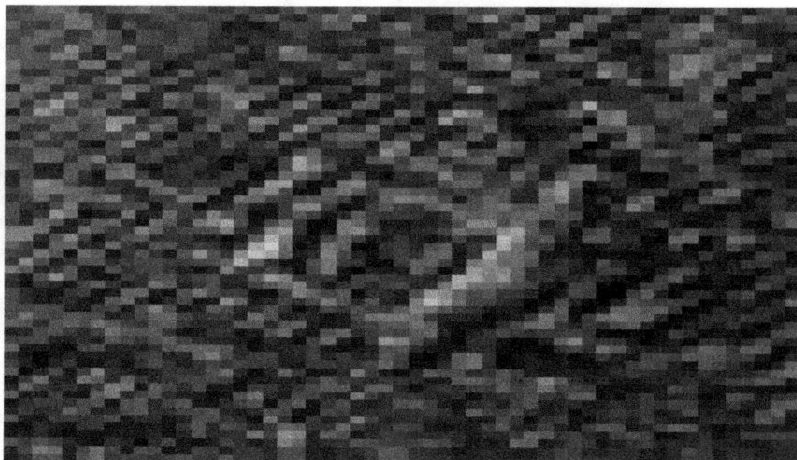

This does not look like a natural formation. NASA's basic explanation for everything is to call it some form of natural formation. But is that always true? Sometimes it is. But Always? Doubtful. Let's make this an example.

Looking at the subject photograph again, one thing that is immediately obvious is the range of features in the landscape itself. It is not one uniform setting, but strikingly different sections of terrain. Note the diamond shapes. They mark a specifically different form of landscape. They do not even account for the unusual white line (pipeline?) passing through them. This is NOT a natural setting. It has been altered by intelligent activity.

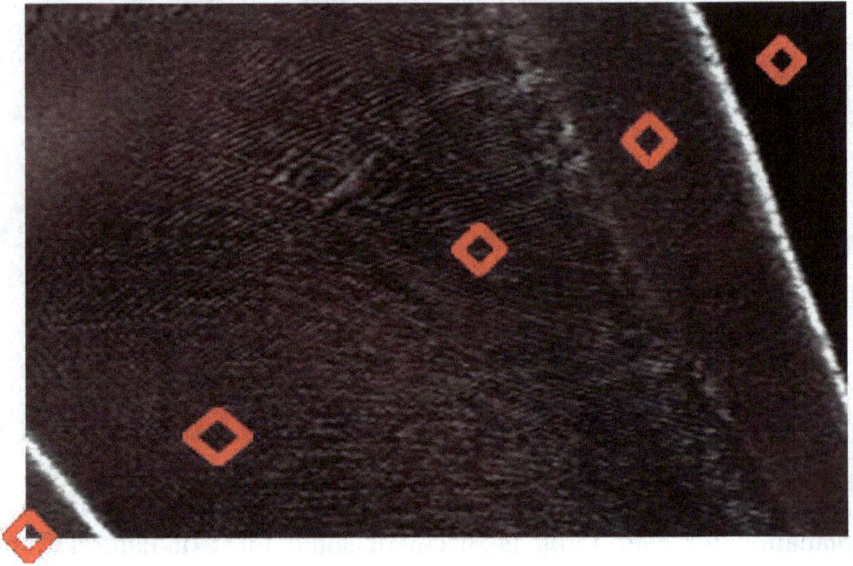

These look more like unnatural features, do they not?

On the subject of terrain, the next example of a Martian anomaly was simply described as cryptic terrain by NASA.

Latitude –81.6, Longitude 120°

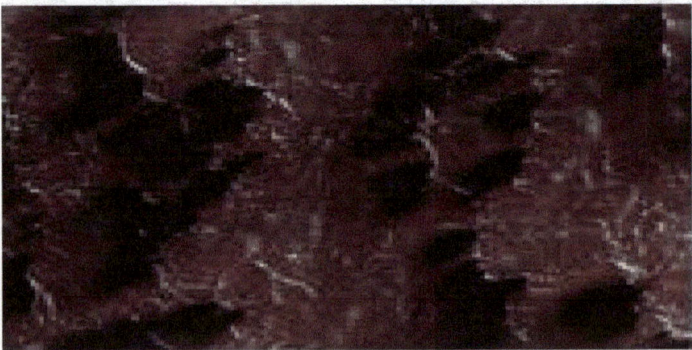

cryptic terrain

132

This appears to be a group of ornamental monuments set up in display.

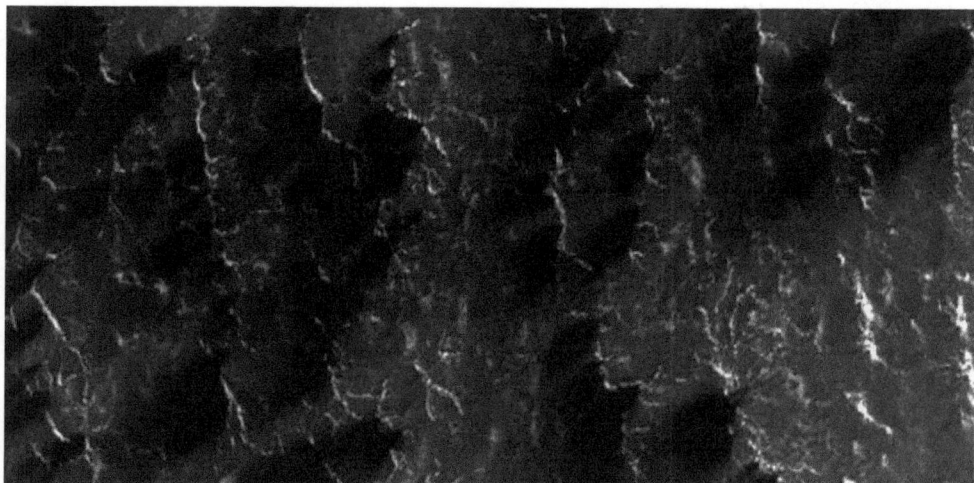

They are very similar to the scene in the next photograph.

Latitude –24.4, Longitude 322.3°

NEAR HOLDEN CRATER

The above almost appear to be arranged in wall–like fashion similar to the Great Wall in China. Below there is another location of wall like structures on Mars.

Latitude 10.6°and Longitude 256°

Considering historical sites on Earth, the next image has a startling resemblance to one on Salisbury Plain in England.

STONEHENGE ON MARS?

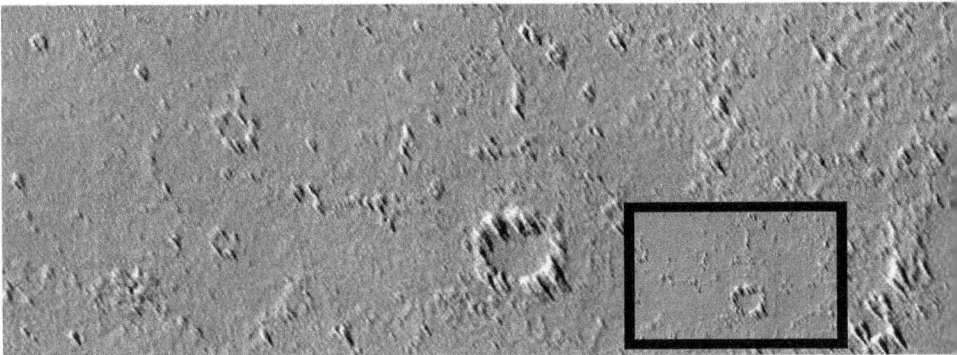

This next example is purely for speculation and a test for one's imagination. This is a bizarre looking terrain, and if one believed in giants on Mars it appears that a fossilized leg bone can be seen here.

Leg bone enclosed in red rectangle.

The flight of imagination continues with the next photograph. This one shows a landscape which projects the profile of a human skull.

skull in Candor Chasma

Latitude −6.8° and Longitude 292.5°

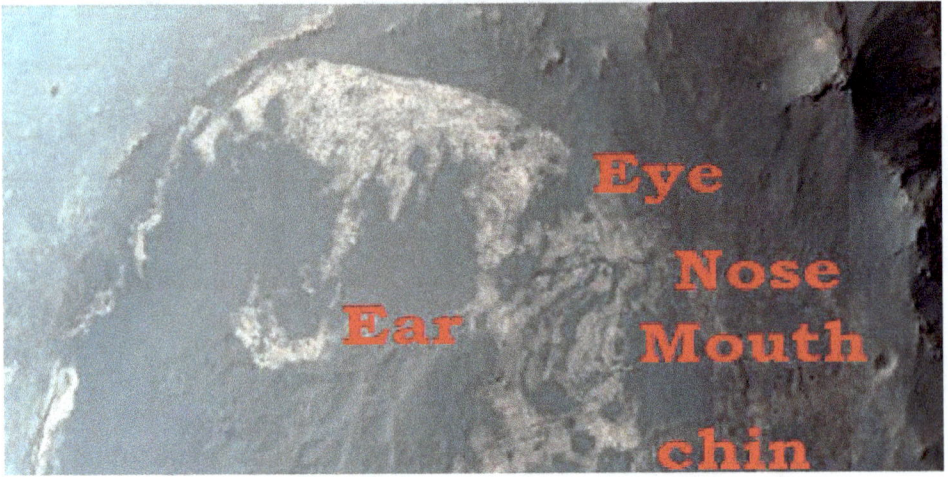

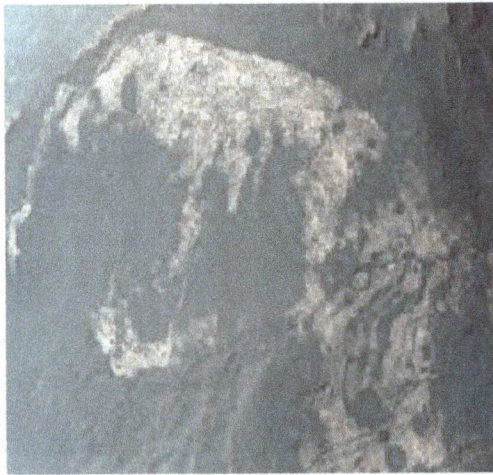

Earlier in the book much consideration was given to potential kiln or manufacturing sites. Below is a fine example of what may be one of these worksites.

latitude 37.4°and Longitude 143.6°

kiln Galaxias Fossae

It may be possible that the surface of Mars had at one time or is currently being used as a type of huge drawing board or chalk board on which an intelligent species is attempting to communicate with the outside universe. The following images make this seem a very possible reality. Could this be an alien language? **Latitude –27.2° by Longitude 328.6°**

Many Small Interesting Ridges in Erythraea Fossa

And from another site:

Where an absolutely stunning discovery is made! A mathematical formula is very clearly seen!

THIS EQUATION WAS TAKEN FROM THE VERY CENTER OF THE FORMATION:

The formula reads (in writing humans can understand): $Y\,3 = t -$ 1. To the right of this are written other figures which I assume are in an as yet unknown alien script.

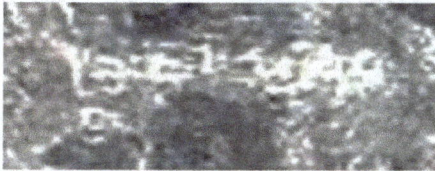

Two different views of the same formulae (earth script and alien script)

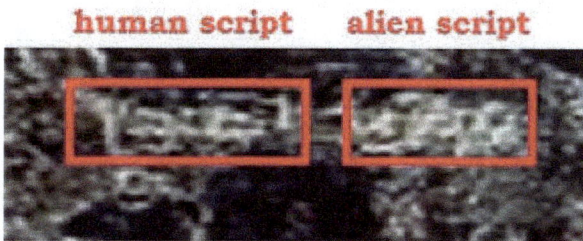

It may be possible to read this formula differently than $Y3 = t - 1$. But it is difficult to read it any other way. It seems quite clear. But what does it mean? Why didn't they use $E = Mc^2$? Maybe because to whoever wrote the equation $Y\,3 = t - 1$ it is more important to them. Maybe it's their definition of the Supreme Being, the Architect of the Universe.

Maybe the equation means: God = t – 1 which symbolizes a great concept to their minds. Who can know right now? But the primary purpose of the equation has fulfilled its goal. It got our attention.

Who created this equation? Maybe the original inhabitants on Mars. The rest of the writing across this astounding section of landscape speaks to its own true existence.

And there is more writing of a different type on the surface of Mars. It is in the form of symbols. Most people are aware that Freemasonry is an international brotherhood which allows all races, creeds and beliefs into its organization. Apparently, it also allows other species into its system as well. What is about to be revealed must make a person ask: was Freemasonry started on Mars? Or was Mars another location on which its great principles were spread by the Architect of the Universe.

Below is the symbol of Freemasonry discovered on the Northern Plains of Mars.

Latitude 51.3 by Longitude 333.9

It may not seem very impressive, but consider the geological processes involved in creating a shape like the one on Mars above. A normal V shape over an inverted V shape. This would require almost impossible tectonics (shifting of underground layers of rock). And it couldn't be accounted for by wind because to create a form such as this would require the wind to form a pattern from both directions at once without dissolving one or the other of the V shapes. This isn't possible. And the form extends over dozens of miles, unbroken.

So where did this symbol come from? Was it scratched out of the soil like the symbols on the Nazca Plain in Peru? Or was it laser cut into the surface by an advanced race? Or is there an even more mysterious origin?

Obviously, this can all be discounted as wishful thinking or simply a wild theory. However, it is more difficult to account for the longitude at which the Masonic symbol was placed: Long: 333.9. It is no secret or mystery that this is one of the most sacred of all masonic numbers. **Is it pure chance, pure coincidence that the symbol on Mars should be placed at this exact location!!**

Considering symbols, another type of completely different one also appears on Mars in a region known for its artistic landscape Asceaus Mons. I call it the hammer and the shield.

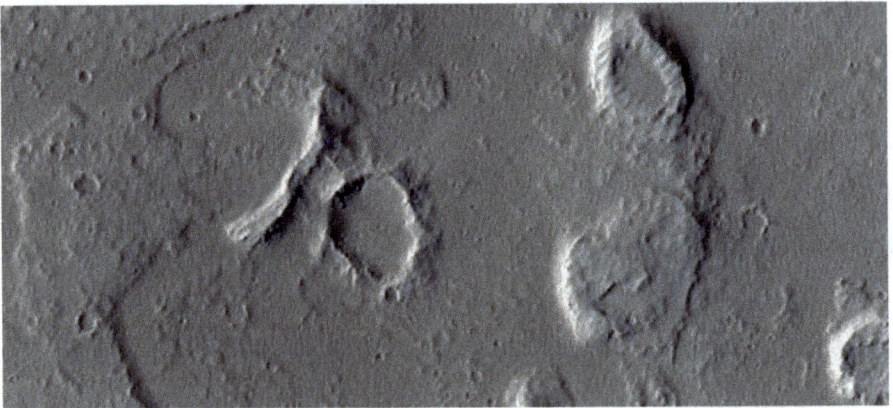

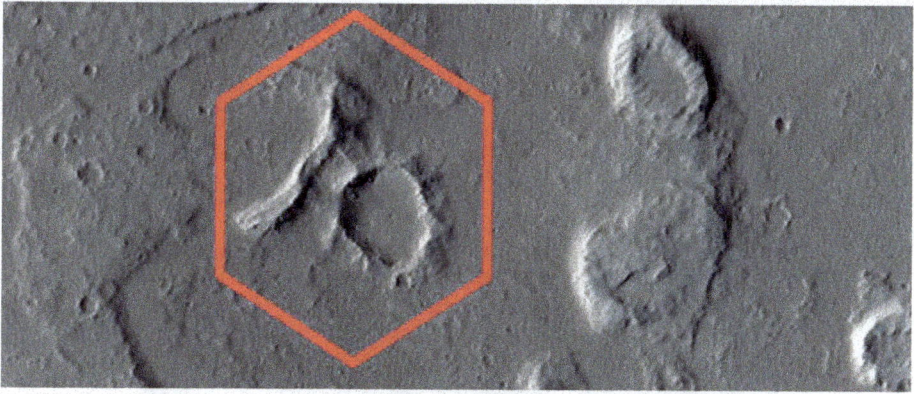

The next image looks very much like a world map, perhaps of the former continents of Mars, which has been created out of surface features.

Melas Chasma 2020 landing site

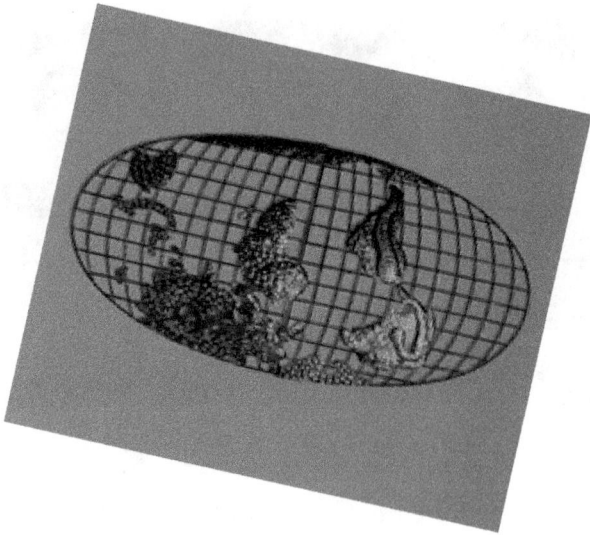

Similar to a wall décor world map.

Latitude –9.8° by Longitude 283.6°

Readers have often asked about the ruins on Mars: where are the highways? The common answer is that either most of them were destroyed during the worldwide catastrophe or else the transportation system the former inhabitants used relied very little on roadways. However, a photograph now exists of a bona–fide Martian highway. It was found in the Ceberus Fosse Region at Latitude 21° and longitude 74.3°.

Highway Ceberus Fossae

What made this photo more impressive was that there was a series of similar images which showed this highway extending onward and onward on a straight path toward the Martian horizon. Unfortunately, they and the above photo have all been deleted – after a single day before the public – from the NASA files. The image did exist. It's before you now! Did NASA suddenly lose it? Below is the message on the NASA page. If there are no images, why was there ever a page 1 for them????

Page 1 of 1 pages (0 images)

Another even more obvious sign of intelligent life on Mars was discovered in Aram Dorsum. An exceptionally bright object was located on the surface here and its origin was of artificial nature.

Aram dorsum

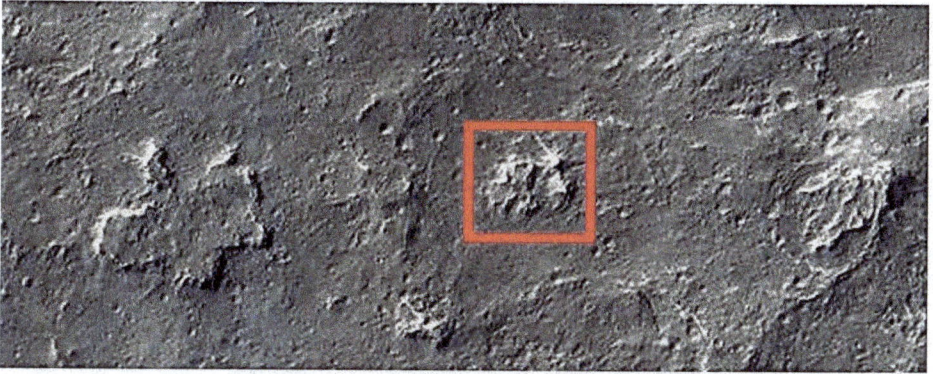

Location of the mysterious light

Close up of the mysterious light

Closer still

It looks like the wreckage of a crashed metallic vehicle. In fact: IT IS! Note the extension to the left. This is clearly a metallic object of artificial design. Is the radiance a distress beacon or simply sun glaring off of the polished metal?

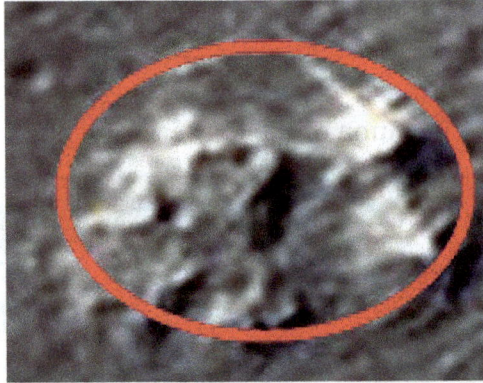

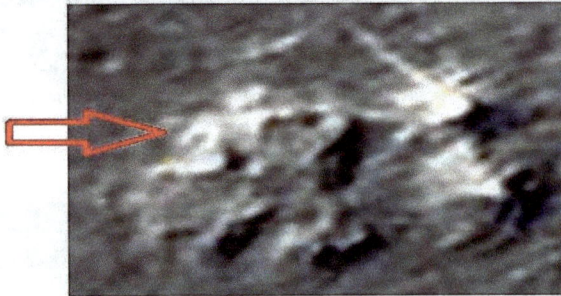

Arrow pointing at piece of warped metal.

Another question often asked is: if there are living beings on Mars why has not one photograph of them ever been taken? Surely even one picture should have been taken even by accident of a living being on Mars by now.

It appears that one finally has been taken. But I will let you be the judge.

Finally, the next image is from a location given by NASA only as the Northern Plains. Unfortunately, there is not any perspective accompanying it either. There isn't any way to determine the size or height of the two "people" or objects in the photo.

At any rate, I would like to view this is a parent out for a stroll across the landscape with a child.

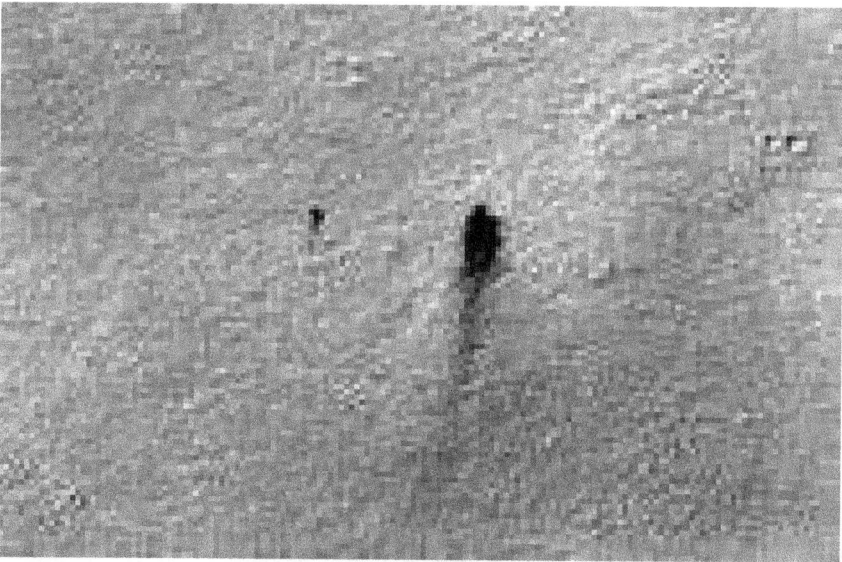

Just a quiet afternoon on Mars.

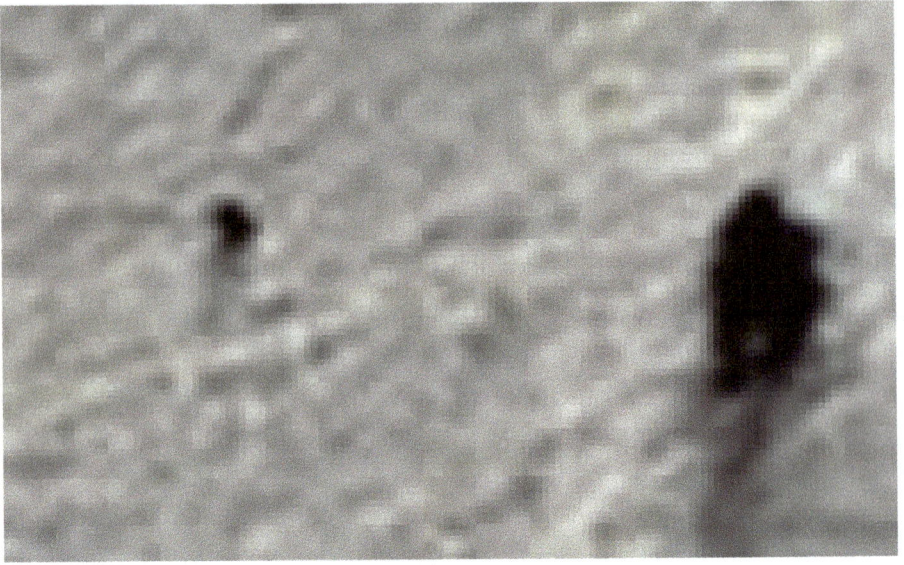

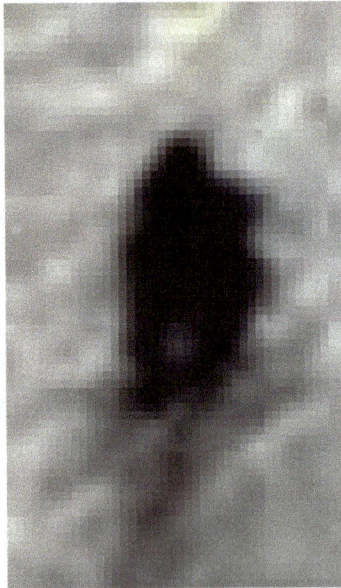

Martian or Sasquatch?

What is most significant about this image is the space that can be seen between the legs of the larger "being." If this were only a mound of rocky material, there would not be a space of this type in this location.

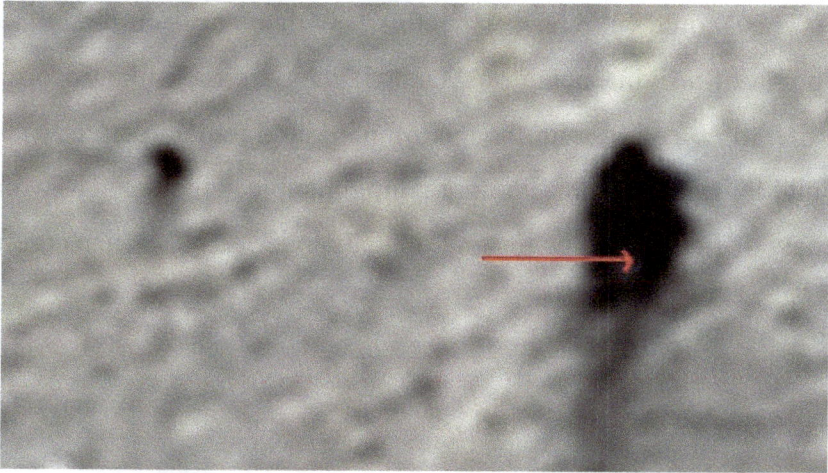

On closer examination, what looks like a pair of feet can also be determined and the "legs" appear to be in motion. Whatever this entity is – be it Sasquatch like or other – it seems to be a verified example of current life on the red planet.

What type of life could it be? I suggest that it is a silicon based life form. Unlike on earth where our life is carbon based, Mars may have developed along a different line. And when you consider the ramifications of this it makes sense. A silicon based life form would be able to endure the high levels of ultraviolet light that pours through the unprotected atmosphere and could also withstand the extremely low barometric pressure that is normal for the red planet. And it would not have to hibernate through the very long winters. So, the entities photographed walking along the surface of Mars may truly be living rock creatures which from a distance bear a basic physical resemblance to our Sasquatches on earth.

This could explain why this is the only currently known photograph of a living being on Mars. If they are rock like of form they could literally blend in with the environment and remain perfectly still for indefinite lengths of time. Photographing a subject such as that would be extremely difficult. Maybe that is how the modern day Martians prefer it!

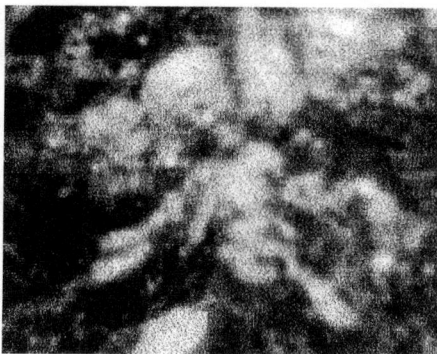

THE END